Cognitive Computing for Smart Automotive Transportation

This reference book explores the integration of cognitive computing technologies in the automotive industry to enhance smart transportation systems. It focuses on how AI, machine learning, and data analytics can improve vehicle automation, safety, and efficiency. Automation can support driverless vehicle transportation and bridge the gap between manual control and fully automated navigation systems. The text introduces a discussion on numerous applications of cognitive computing in smart transportation, motion planning, situation awareness, dynamic driving, adaptive behavior, human intent measurement, and predictive analysis.

Key Features:

- Discusses basic concepts and architecture of cognitive computing for vehicular systems.
- Presents technologies to measure human intent for vehicle safety, including emergency management systems (EMS).
- Covers the perception and localization processes in autonomous driving through LiDAR, GPS, and Stereo vision data with critical decision-making and simulation results.
- Elucidates the application of motion planning for smart transportation.
- Covers visual perception technologies for advanced driver assistance systems (ADAS) through deep learning.

The text is primarily written for graduate students, academic researchers, and professionals in the fields of computer science, electrical engineering, automotive engineering, and civil engineering.

Chapman & Hall/CRC Internet of Things: Data-Centric Intelligent Computing, Informatics, and Communication

The role of adaptation, machine learning, computational intelligence, and data analytics in the field of IoT systems is becoming increasingly essential and intertwined. The capability of an intelligent system is growing depending upon various self-decision-making algorithms in IoT devices. IoT-based smart systems generate a large amount of data that cannot be processed by traditional data processing algorithms and applications. Hence, this book series involves different computational methods incorporated within the system with the help of analytics reasoning, learning methods, artificial intelligence, and sense-making in big data, which are most concerned in IoT-enabled environments.

This series focuses on attracting researchers and practitioners who are working in information technology and computer science in the fields of intelligent computing paradigm, big data, machine learning, sensor data, Internet of Things, and data sciences. The main aim of the series is to make available a range of books on all aspects of learning, analytics, and advanced intelligent systems and related technologies. This series will cover the theory, research, development, and applications of learning, computational analytics, data processing, and machine learning algorithms, as embedded in the fields of engineering, computer science, and information technology.

Series Editors: Souvik Pal

Energy Harvesting: Enabling IoT Transformations
Deepti Agarwal, Kimmi Verma, and Shabana Urooj

SDN-Supported Edge-Cloud Interplay for Next Generation Internet of Things
Kshira Sagar Sahoo, Arun Solanki, Sambit Kumar Mishra, Bibhudatta Sahoo, and Anand Nayyar

Internet of Things: Applications for Sustainable Development
Niranjan Lal, Shamimul Qamar, Sanyam Agarwal, Ambuj Kumar Agarwal, and Sourabh Singh Verma

Artificial Intelligence for Cognitive Modeling: Theory and Practice
Pijush Dutta, Souvik Pal, Asok Kumar, and Korhan Cengiz

Cognitive Predictive Maintenance Tools in Brain Diseases: Design and Analysis
Shweta Gupta

Assimilation of Blockchain Technology with IoT: Industry Perspectives
R Anandan, D Akila, and Souvik Pal

Cognitive Computing for Smart Automotive Transportation: Technology and Applications
By G. R. Kanagachidambaresan, Archana Naganathan, Niresh Jayarajan, and Sheldon S. Williamson

Cognitive Computing for Smart Automotive Transportation

Technology and Applications

Edited by

G. R. Kanagachidambaresan, Archana Naganathan, Niresh Jayarajan, and Sheldon S. Williamson

CRC Press
Taylor & Francis Group
Boca Raton London New York

CRC Press is an imprint of the
Taylor & Francis Group, an **informa** business
A CHAPMAN & HALL BOOK

Designed cover image: ShutterStock
First edition published 2025
by CRC Press
2385 NW Executive Center Drive, Suite 320, Boca Raton FL 33431

and by CRC Press
4 Park Square, Milton Park, Abingdon, Oxon, OX14 4RN

CRC Press is an imprint of Taylor & Francis Group, LLC

ISBN: 9781032487403 (hbk)
ISBN: 9781032491455 (pbk)
ISBN: 9781003392330 (ebk)

DOI: 10.1201/9781003392330

Typeset in Times
by codeMantra

To students, scholars, friends, and family members.

Contents

Preface

Automation has the potential to facilitate the transportation of driverless vehicles and to serve as a bridge between completely automated navigation systems and manual control. Cognitive computing offers a practical approach to artificial intelligence. These self-learning systems evolve into autonomous systems that are independent of human interventions by utilizing machine learning to emulate the human brain, thereby facilitating automation in a variety of application areas. The most recent advancements in the automotive industry are autonomous driving vehicles or connected automobiles. The introduction of cognitive computing technology has enabled the development of solutions that are on par with human intelligence for both in-vehicle intelligent systems and off-road vehicle diagnosis and maintenance. This is due to the self-learning and self-healing capabilities of cognitive computing technology, which have been employed to enhance vehicle safety. We are confident that these cognitive computing models will have a substantial future impact by resolving intricate problems with ambiguous solutions.

This edited volume examines the security features required in smart transportation, the role of cognitive computing in smart mobility, and a variety of applications of cognitive computing in the context of smart automotive transportation. This book also discusses the role of connected cars and fault detection mechanisms; the design of internal combustion (IC) engines for smart cars with high-performance computing; and the fundamental concepts and architecture of cognitive computing, motion planning, situation awareness, dynamic driving, adaptive behavior, human intent measurement, and predictive analysis.

This book is designed for a diverse audience, including advanced students, researchers, and industry practitioners. This resource will be beneficial for individuals who intend to conduct research in the field of cognitive computing, specifically in the context of autonomous and intelligent driving vehicles.

About the Authors

G. R. Kanagachidambaresan is a professor in the Department of Computer Science and Engineering at Vel Tech Rangarajan Dr. Sagunthala R&D Institute of Science and Technology. He earned his PhD from Anna University in 2017. Presently, he is handling funded projects from ISRO, DBT, and DRDO. He has authored books and articles in the area of IoT, wireless networks, and expert systems. He has worked as a consultant for several leading MNC companies. He is the managing director of Eazythings Technology Private Limited and a TEC committee member of DBT. He is also the editor in chief of the *Next Generation Computing and Communication Engineering* series.

Archana Naganathan has been an assistant professor in the Department of Electrical and Electronics Engineering, PSG College of Technology, Coimbatore, India, for the past 13 years. Her research interests cover the areas of power electronics, optimization techniques, and hybrid electric vehicles. She has received a travel grant from DST for an international conference and has completed a DST project on a deep learning-based animal intrusion detection system with a grant of 30 lakhs. Currently, she is working on an IIT-IPTIF project in the area of Edge AI. She has published a patent in the area of power converters for electric vehicle charging and has published around 40 papers in various journals and conferences. She has authored 4 books. She earned a PhD degree from Anna University, Chennai, India, in 2017.

Niresh Jayarajan holds a doctorate degree in the field of optimization and has been with PSG College of Technology since 2011. His areas of research include electric and hybrid vehicles, automotive electronics and optimization by using soft computing techniques, and unmanned aerial vehicles (UAVs). He is guiding doctoral scholars in the area of UAVs and electric vehicle. He serves as a reviewer for prestigious journals including *Elsevier* and *Springer* publication. He has published around

58 international/national journals and around 7 national/international conferences in NIT/IIT institutes. In accordance, he has also published 7 books under a reputed publishing banner.

Sheldon S. Williamson (Fellow, IEEE) earned a B.E. (Hons.) degree in electrical engineering from the University of Mumbai, Mumbai, India, in 1999. He earned M.S. and Ph.D. (Hons.) degrees in electrical engineering from the Illinois Institute of Technology, Chicago, IL, USA, in 2002 and 2006, respectively. He is a professor with the Department of Electrical, Computer, and Software Engineering and the director of Smart Transportation Electrification and Energy Research (STEER) Group, Faculty of Engineering and Applied Sciences, Ontario Tech University, Oshawa, ON, Canada. His research interests include advanced power electronics, electric energy storage systems, and motor drives for transportation electrification. He holds the prestigious NSERC Canada Research Chair position in electric energy storage systems for transportation electrification.

List of Contributors

B. Amutha
Faculty of Engineering and Technology, Department of Computing Technologies, School of Computing
SRM Institute of Science and Technology
Kattankulathur (Chengalpattu District), Tamil Nadu, India

M. P. Anbarasi
Department of Robotics & Automation
PSG College of Technology
Coimbatore, Tamil Nadu, India

Anjali Arora
Department of Computer Science & Engineering
Shri Ram Murti Smarak College of Engineering and Technology
Bareilly, Uttar Pradesh, India

S. Delsi Robinsha
Faculty of Engineering and Technology, Department of Computing Technologies, School of Computing
SRM Institute of Science and Technology
Kattankulathur (Chengalpattu District), Tamil Nadu, India

G. Gowtham
Department of Electronics and Communication Engineering
Kumaraguru College of Technology
Coimbatore, Tamil Nadu, India

Roshni Kapoor
Department of Computer Science & Engineering
Shri Ram Murti Smarak College of Engineering and Technology
Bareilly, Uttar Pradesh, India

Benjamin Lawler
Department of Automotive Engineering
Clemson University
Clemson, South Carolina

N. Murugu Nachippan
Department of Automobile Engineering
Easwari Engineering College
Chennai, Tamil Nadu, India

J. Niresh
Department of Automobile Engineering
PSG College of Technology
Coimbatore, Tamil Nadu, India

Avinash Ravikumar
Department of Automotive Engineering
Clemson University
Clemson, South Carolina

Pradeep Kumar Sharma
Department of Computer Science & Engineering
Shri Ram Murti Smarak College of Engineering and Technology
Bareilly, Uttar Pradesh, India

C. Shri Raghavi
Department of Robotics & Automation
PSG College of Technology
Coimbatore, Tamil Nadu, India

P. Shudhi Rishaa
Department of Robotics & Automation
PSG College of Technology
Coimbatore, Tamil Nadu, India

S. Umamaheswari
Department of Electronics and Communication Engineering
Kumaraguru College of Technology
Coimbatore, Tamil Nadu, India

D. Vanusha
Faculty of Engineering and Technology, Department of Computing Technologies, School of Computing
SRM Institute of Science and Technology
Kattankulathur (Chengalpattu District), Tamil Nadu, India

Balaji Vasudevan
Department of Mechanical Engineering
Vel Tech Rangarajan Dr. Sagunthala R&D Institute of Science and Technology
Chennai, Tamil Nadu, India

D. Vathana
Faculty of Engineering and Technology, Department of Computing Technologies, School of Computing
SRM Institute of Science and Technology
Kattankulathur (Chengalpattu District), Tamil Nadu, India

M.d. Wazih
Department of Computer Science & Engineering
Shri Ram Murti Smarak College of Engineering and Technology
Bareilly, Uttar Pradesh, India

1 Research on Security and Privacy Issues in IoV-Enabled Smart Transportation Systems

Anjali Arora, Roshni Kapoor, M.d. Wazih, and Pradeep Kumar Sharma

1.1 INTRODUCTION

1.1.1 Introduction to IoV-Enabled Intelligent Transportation Systems

The development of intelligent transportation systems (ITS) is changing how our cities and roads operate and what's possible to drive on. Global travel efficiency, safety, and overall experience could all be enhanced with ITS [1]. Prominent organizations and localities are launching campaigns to support information and communication technology breakthroughs. Examples include British Telecom's CityVerve project and IBM's Smarter City program. The transportation scene is set to undergo a dramatic shift due to the rise of autonomous or semi-autonomous vehicles and powerful traffic forecasting systems. Furthermore, modern conveniences are being easily integrated into cars thanks to developments in car entertainment systems, which include Bluetooth connectivity, hands-free communication, and navigation aids. The term "Internet of Vehicles" (IoV) refers to this networked technology advancement in transportation that integrates automobiles, communication networks, and a range of features. The combination of various technologies is said as (IoV). The IoV integrates vehicles, communication networks, and various transportation infrastructure elements into a smart and interconnected system [2]. With advanced technologies, it enhances safety, efficiency, and overall mobility in urban and rural environments. The IoV framework is a mind-boggling system made out of various organization members, for example, vehicles, roadside units (RSUs), and individuals, imparting through remote networks. It is made up of IoV, which are frequently outfitted with radar, GPS, vision sensors, and other technologies. They can gather street climate information and constant traffic conditions and submit them to the cloud so that different users can design their courses ahead of time by getting to this information. Users who share data might confront extreme security and protection risks, like spillage of driver personality, current location, and navigation information. Simultaneously, due to high liquidity and unpredictability, in-vehicle networks are powerless to different dangers [3].

DOI: 10.1201/9781003392330-1

1.1.2 Overview of the Internet of Vehicles (IoV)

The IoV is a complex organization that utilizes sensors, programming, and innovation to interface vehicles with their surroundings, such as other vehicles, pedestrians, traffic signals, and parking facilities, allowing them to share data. IoV is an evolution in transportation and traffic management systems, enabling seamless communication between cars and infrastructure. It is built on the foundation of the Internet of Things (IoT) and vehicular ad hoc networks (VANETs) [4].

The future automated vehicles will have numerous associated end points and a large volume of information shared. Later on, any vehicle will be able to connect to anything at any time in a really adaptive, strong, and secure manner. 5G communication will be critical for these network connectivity requirements in order to supply the necessary services. 5G is generally suited for car-to-infrastructure networks due to its extremely low latency, high communication speed, and accessibility to a large volume of information interchange [5]. The following are some of the key advantages provided by IoV in terms of improving overall transportation infrastructure:

1. **Reduced Traffic Congestion**: IoV can optimize traffic patterns and alleviate congestion by using vehicle data from sensors and real-time traffic monitoring.
2. **Smart Parking Solutions**: IoV provides smart parking solutions in densely populated and congested metropolitan areas, reducing traffic congestion and enhancing city traffic management.
3. **Vehicle Maintenance**: IoV allows automakers to use data to discover product faults, while sensors offer drivers with regular updates on the status of their vehicle, decreasing breakdowns and accidents.
4. **Impact on Energy Conservation and Sustainability**: IoV, when paired with Edge artificial intelligence and analytics, dramatically contributes to energy savings and reduces cities' overall carbon footprint.

The applications of IoV go beyond automobiles to include the entire traffic and transportation ecosystem. IoV can optimize fuel and energy resources by adding intelligent signals and lighting that monitor traffic conditions and change accordingly. Furthermore, IoV technology has the potential to greatly assist the environment by lowering carbon emissions through improved vehicle management and maintenance procedures. Connected automobiles significantly contribute to decreasing carbon emissions and enhancing global ecological sustainability. However, like with every technological innovation, IoV has its concerns, particularly in terms of security and protection, which should be tended to completely. To tackle these issues, the government is initiating programs and allocating funds to projects aimed at enhancing the effectiveness and efficiency of ITS [6,7].

1.1.3 Importance of Security and Protection in Smart Transportation

Security and protection are essential aspects in the design and implementation of ITS. As technology continues to advance, intelligent transportation solutions, such as connected vehicles, intelligent traffic management systems, and autonomous vehicles, provide a host of advantages; however, they also bring forth new challenges and risks [8,9]. Here are several reasons highlighting the importance of security and privacy in smart transportation:

1. **Safety Concerns**: Security breaches in smart transportation systems can have direct consequences on safety. For example, unauthorized access to connected vehicles could lead to malicious activities like tampering with critical systems or taking control of the vehicle.
2. **Data Integrity**: Intelligent transportation is primarily dependent on data, which include real-time traffic statistics, vehicle status, and navigation details. Ensuring the integrity of these data is essential for making well-informed decisions and upholding the reliability of the transportation system.
3. **Personal Privacy**: Connected vehicles and transportation infrastructure generate massive amounts of data, which frequently include personal information. Individual privacy must be protected by safeguarding these data. Illegitimate entry into personal information may lead to identity theft, individual surveillance, and various privacy concerns.
4. **Resilience to Cyberattacks**: Cyberattacks on ITS have the potential to disrupt routine operations, cause accidents, or compromise the efficiency of the transportation network. Building resilience to such attacks is vital to maintain the reliability and functionality with regard to the transportation infrastructure.
5. **Public Trust and Adoption**: The public's trust is essential for the successful implementation ofITS. Apprehensions regarding security and privacy may discourage individuals from utilizing these systems. Ensuring robust security measures and privacy protection mechanisms can enhance public confidence in smart transportation solutions.
6. **Legal and Regulatory Compliance**: Governments and regulatory agencies are becoming more aware of the significance of security and privacy in smart transportation. Compliance with legal and regulatory frameworks helps prevent legal issues and ensures the implementation of smart transportation technologies that are compatible with ethical and legal standards.
7. **System Interoperability**: A variety of interconnected components and devices are used in intelligent transportation. Security measures must be implemented together to ensure interoperability and secure communication among the various components of the transportation ecosystem.
8. **Protection against Physical and Cyber Threats**: Smart transportation systems are vulnerable to both physical and cyber threats. They include not only cyberattacks on software and communication networks but also physical attacks on infrastructure. Security measures are essential to protect against these diverse threats.
9. **Data Sharing and Collaboration**: Collaborative efforts and data sharing among different entities in the transportation ecosystem can lead to more efficient and effective systems. However, this must be done with a strong emphasis on security and privacy to prevent unauthorized access and misuse of sensitive information.

1.2 IoV ARCHITECTURE AND ITS COMPONENTS

In the realm of smart transportation, system architecture typically involves the design and organization of various components to enable efficient and secure communication among vehicles, infrastructure, and other entities. The architecture [10,11] may include elements such as:

1. **Vehicle-to-Everything (V2X) Communication:** This incorporates communication between vehicles and different elements, including vehicle-to-infrastructure (V2I), vehicle-to-pedestrian (V2P), and vehicle-to-vehicle (V2V) communications.
2. **Data Processing and Analytics:** The architecture might include systems for processing and analyzing the extensive data produced by connected vehicles. This could entail utilizing cloud-based solutions, edge computing, or a hybrid combination of both.
3. **Security Measures:** Considering the sensitivity of transportation systems, security emerges as a crucial aspect. The architecture should include robust security measures to protect data, communication channels, and the overall system from cyber threats.
4. **Sensors and IoT Devices:** Smart transportation systems rely on various sensors and IoT devices on vehicles and infrastructure to collect data related to traffic conditions, road quality, and other relevant parameters.
5. **Integration with Urban Infrastructure:** The architecture may consider integration with urban infrastructure elements, including traffic lights and road signs, and smart city systems, to optimize traffic flow and enhance safety.
6. **User Interfaces and Applications:** For end-users, there may be applications or user interfaces that provide real-time information, navigation assistance, and other services.

1.2.1 Communication Protocols in IoV

In the context of IoT in smart transportation, various communication protocols are used for facilitating communication between vehicles and infrastructure. Some of the frequently employed communication protocols in the context of the IoV [12,13] for smart transportation encompass the following:

- **Dedicated Short-Range Communications (DSRC)**: DSRC is a wireless communication protocol specifically designed for vehicular communication. It works in the 5.9 GHz frequency range and provides low-latency, high-speed communication among vehicles and roadside infrastructure. DSRC is generally utilized for applications such as V2V communication, V2I communication, and V2P communication in applications like ITS and connected vehicle environments (Figure 1.1).
- **Cellular Vehicle-to-Everything (C-V2X)**: C-V2X stands as a nascent communication technology that utilizes cellular networks for vehicle communication. It is designed to support both low-latency, short-range communication and wide-area communication (communication with cellular

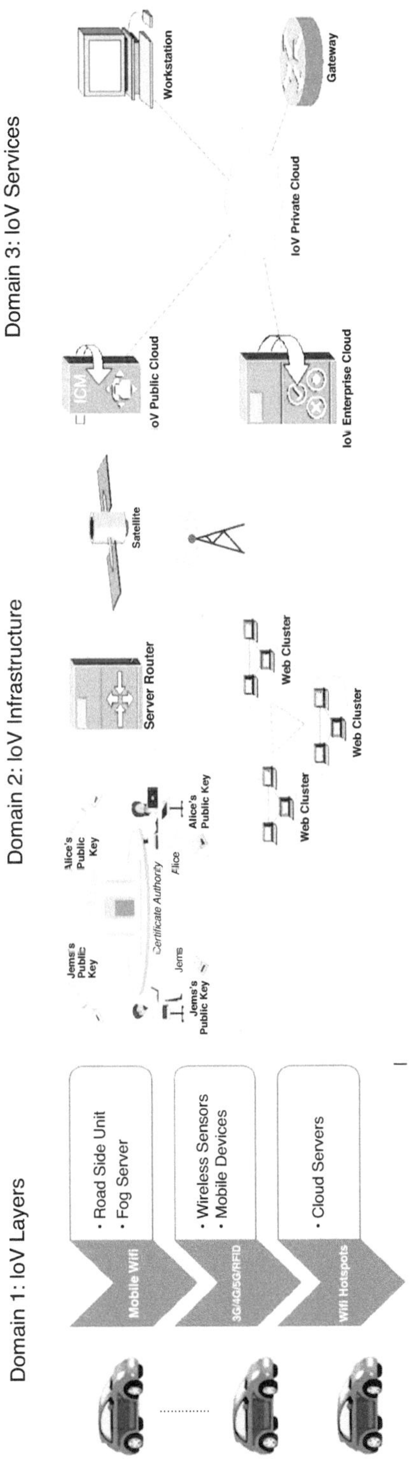

FIGURE 1.1 IoV architecture.

infrastructure). C-V2X operates in licensed cellular bands and can leverage existing LTE and 5G networks. C-V2X offers enhanced safety, traffic efficiency, and infotainment services in smart transportation (Figure 1.2).

- **Wi-Fi (IEEE 802.11p)**: IEEE 802.11p, ordinarily alluded to as remote access in vehicular conditions, represents a Wi-Fi variant uniquely crafted for vehicular communication. It works inside the 5.9 GHz frequency range, delivering low-latency communication for V2V and V2I applications. IEEE 802.11p finds extensive use in applications like collision avoidance, traffic signal synchronization, and cooperative driving.
- **Vehicle-to-Infrastructure (V2I) Protocols**: Various protocols, often based on standard Internet protocols (such as IPv6), are used for communication between vehicles and infrastructure. These protocols enable vehicles to exchange information with traffic management systems, roadside units, and ITS. Examples of V2I protocols include simple network management protocol (SNMP), media-independent handover (MIH) protocols, and session initiation protocol (SIP).
- **Message Queuing Telemetry Transport (MQTT)**: MQTT is a lightweight messaging protocol based on the publish/subscribe model, broadly utilized in IoT applications, including smart transportation. It guarantees efficient and reliable communication among vehicles, sensors, and cloud platforms. MQTT is suitable for low-power and low-bandwidth networks and enables real-time data exchange for applications like remote monitoring, traffic management, and fleet management.

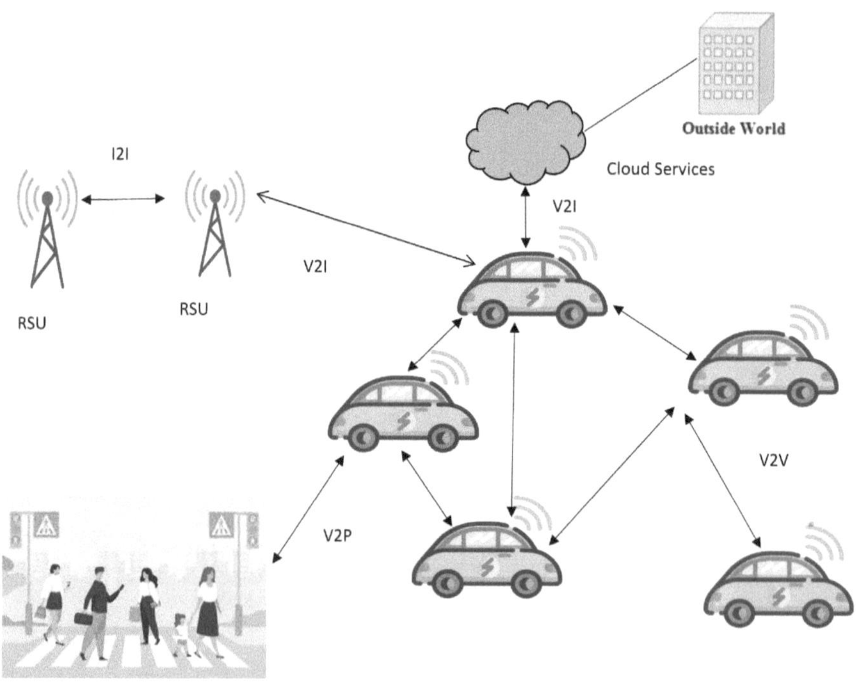

FIGURE 1.2 IoV communication.

- **Cooperative Intelligent Transport Systems (C-ITS) Protocols**: C-ITS protocols are standardized communication protocols used for ITS to support cooperative driving and infrastructure-to-vehicle communication. These protocols enable vehicles to exchange safety-related messages, such as cooperative awareness messages (CAM) and decentralized environmental notification messages (DENM). Examples of C-ITS protocols include ITS-G5, C-ITS application layer messages (CALM), and ETSI ITS-G5 protocol suite.

It's crucial to emphasize that communication protocols used in IoV may differ based on the region, regulatory frameworks, and infrastructure availability. Additionally, advancements in cellular networks, such as the rollout of 5G, are anticipated to have a substantial impact in enabling advanced communication capabilities in smart transportation [14].

1.2.2 V2X Communication

V2X technology [15,16] involves utilizing ITS communication networks to transmit crucial information between vehicles and various components of the transportation infrastructure. This includes relaying safety alerts, weather updates, traffic conditions, and route data both from vehicles to the infrastructure and vice versa. V2X encompasses two primary aspects: V2V and V2I. In V2V, vehicles interact akin to VANETs, exchanging data on speed, position, and potential road hazards. V2I communication facilitates bidirectional wireless data exchange between vehicles and the road infrastructure. This includes elements like RFID readers, cameras, lane markings, traffic signals, streetlights, road signs, and parking meters. Smart vehicles outfitted with on-board units (OBUs) layout communication with RSUs, nearby IoT devices, and other OBUs as they traverse roadways. Through the V2X framework, a spectrum of signals and messages is transmitted and received to enhance roadway efficiency and safety for all travelers. However, while V2X bolsters system connectivity, it also introduces a new vulnerability that could be exploited by malicious entities [17].

1.3 THREAT LANDSCAPE IN IoV

The three primary points within ITS where security techniques must be applied include data collection, remote information transmission over the organization, and ensuing information investigation. This section discusses security measures to guarantee that the ITS will not be vulnerable to the aforementioned threats [18] (Figure 1.3).

1. **Reliability**: Employing confidentiality measures like encipherment, among others, is essential for guaranteeing confidentiality and secrecy of information conveyed. Data collected by various sensors and Internet of Things (IoT) devices placed at various points of interest must be kept private. Such sensitive data can be given a complex security model through the use of visual techniques [19].
2. **Authentication**: Ensuring robust device identity mechanisms can provide both entity authentication and channel authentication. To maintain authentication, it is imperative to prevent the Illicit takeover of IoT devices or manipulation of sensors and other devices [20].

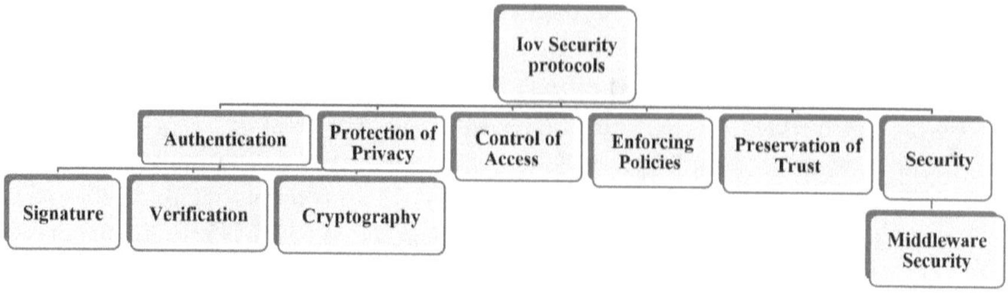

FIGURE 1.3 IoV security protocols.

3. **Control of Access**: In order to guarantee legitimate access to devices and data, as well as effective access control, policies and procedures need to be established. To uphold the security and assurance of information, as well as the confidentiality of other participating individuals, it is crucial to establish a hierarchical access control mechanism. This mechanism assigns distinct privileges for access to individuals [21].
4. **Protection of Privacy**: Particularly in this kind of network, privacy is the most crucial issue. Ensuring adherence to the regulations of the specified geographic area will protect privacy. At no time should privacy be sacrificed for computational efficiency. On the other hand, precise traffic recommendations occasionally have the potential to partially violate privacy protection [22].
5. **Enforcing Policies**: The regulatory body must enforce proper norms and compliance in order to establish communication among diverse automobiles.
6. **Preservation of Trust**: Once communication commences, privacy cannot be regulated due to device heterogeneity. Establishing trust between the communicating entities before initiating communication is one of the reasons behind this. To enhance the reliability of network communication, the trust management paradigm is employed to prevent security threats and attacks. Additionally, it improves the network's resilience and the effectiveness of computational resources [23].
7. **Middleware Used to Be Secure**: The middleware employed for data collection, transportation, and analysis must be safeguarded against manipulation, hardware malfunctions, exploitation-related damages, and potential hijacking. Any security strategy implemented in ITS should be durable enough to support future computing requirements [24].

1.3.1 Cybersecurity Threats in Smart Transportation

Smart transportation systems face various cybersecurity threats that can compromise safety, data integrity, and operational continuity:

- **Unauthorized Access and Control**: Hackers may attempt to gain unauthorized access to control systems within smart transportation infrastructure. This could include traffic control systems, autonomous vehicle controls, or even infrastructure management systems, allowing them to manipulate operations or cause accidents.

- **Ransomware Attacks**: Malicious actors may deploy ransomware to encrypt critical systems or data, and attackers demand payment to restore access. In smart transportation, this might lead to service disruptions, rendering systems non-operational until the ransom is paid.
- **Denial-of-Service (DoS) Attacks**: Attackers might flood transportation systems with excessive traffic and overwhelming networks and cause services to become unavailable. This could affect traffic management systems, leading to congestion or disruption of essential services.
- **Vulnerabilities in Connected Vehicles**: With the increasing integration of IoT devices in vehicles, software vulnerabilities or hardware can be taken into account. Hackers may obtain entry to vehicle systems, compromising safety features or stealing sensitive information.
- **Supply Chain Attacks**: Weaknesses in third-party software or components used in smart transportation systems can be used by adversaries to enter the infrastructure uninvited or introduce malicious programs.

To moderate these dangers, it is imperative to execute robust cybersecurity measures such as encryption, network segmentation, regular security audits, intrusion detection systems, access controls, and employee training. Additionally, adopting a proactive approach to identifying and patching vulnerabilities in both hardware and software components is essential to strengthen the resilience of smart transportation systems against cyber threats [25,26].

1.3.2 Attacks on IoV Infrastructure and Communication

The IoV holds immense potential for enhancing transportation safety, efficiency, and comfort. However, this interconnectedness also exposes the system to various security threats that can potentially put passengers, drivers, and infrastructure at risk [27]. Let's dive into the main categories of these threats.

1. Inter-Vehicle Attacks:
    - **Spoofing**: Attackers can imitate legitimate vehicles or infrastructure elements to spread misinformation, manipulate traffic flow, or even cause accidents. Examples include:
    - **GPS Spoofing**: Manipulating a vehicle's GPS data to disrupt navigation or trigger false warnings.
    - **Identity Spoofing**: Pretending to be another vehicle or trusted entity to gain unauthorized access or control.
    - **Denial-of-Service (DoS)**: Flooding the network with fake data to overwhelm and disable communication channels, hindering emergency services or traffic management.
    - **Data Eavesdropping**: Intercepting sensitive data traded among vehicles and infrastructure, including location data, driving behavior, or sensor readings, poses a threat. These data can be taken advantage of for different malignant purposes, such as profiling drivers, blackmail, or targeted attacks [28].

2. Intra-Vehicle Attacks:
    - **Software Vulnerabilities**: Exploiting weaknesses in vehicle software or firmware to gain control over critical systems like brakes, steering, or engine management. This can lead to catastrophic consequences like remote hijacking or system malfunction.
    - **Man-in-the-Middle Attacks**: Intercepting and manipulating communication between different internal components of a vehicle, potentially causing malfunction or disrupting essential functions.
    - **Physical Tampering**: Gaining direct access to the vehicle's hardware to install malware or manipulate sensors, posing a serious threat to safety and privacy [29].
3. Other Threats:
    - **Data Privacy Concerns**: The extensive volume of data collected by IoV systems prompts worries regarding potential misuse and unauthorized access. Robust data privacy regulations and strong encryption are crucial to protecting user information.
    - **Infrastructure Vulnerabilities**: Attacks on roadside infrastructure, like traffic lights or communication towers, can disrupt traffic flow and compromise overall safety.

    Addressing these threats necessitates a multi-layered approach, encompassing:
    - **Robust Security Protocols and Encryption**: Enforcing robust authentication, authorization, and encryption methods to safeguard communication channels and data integrity.
    - **Software Updates and Vulnerability Patching**: Regularly updating vehicle software and firmware to address vulnerabilities and prevent exploitation.
    - **Threat Intelligence and Intrusion Detection Systems**: Employing advanced security solutions to monitor IoV networks for suspicious activity and proactively detect potential threats.

Securing the IoV ecosystem is crucial for realizing its full potential and ensuring the safety, privacy, and trust of all participants. By tackling these challenges and implementing robust security measures, we can help lay the foundation for a transportation future that is safer and more efficient [30].

1.3.2.1 Risks Associated with Connected Vehicles

Connected vehicles, which are equipped with internet connectivity and communication technologies, offer various benefits such as improved safety, efficiency, and convenience. However, they also come with several risks and difficulties [31]. Here are some of the key risks associated with connected vehicles:

1. Cybersecurity Threats:
    - **Hacking and Unauthorized Access**: Connected vehicles are susceptible to cyberattacks, including hacking attempts by malevolent parties attempting to obtain unapproved access to vehicle systems. This could lead to theft, privacy breaches, or even control over critical vehicle functions.

- **Data Breaches**: The substantial volume of data produced by connected vehicles, including location information, driving patterns, and personal preferences, can be a target for cybercriminals. A data breach can have severe consequences for user privacy.

2. Privacy Concerns:
   - **Location Tracking**: Connected vehicles collect and transmit location data, raising concerns about constant tracking and surveillance. Unauthorized access to this information can compromise user privacy.
   - **Personal Data Exposure**: As vehicles become more connected, there is an increased risk of exposure of personal information. This may include driver profiles, contact details, and usage patterns.
3. Safety Risks:
   - **Malicious Control**: If a connected vehicle's systems are compromised, malicious actors could gain control over critical functions, leading to unsafe driving conditions and accidents.
   - **System Failures**: Connectivity introduces the risk of system failures or glitches that could impact the vehicle's performance, leading to accidents or malfunctions.
4. Infrastructure Vulnerabilities:
   - **Communication Network Risks**: The reliability and security of the communication networks that connected vehicles rely on can be compromised. This may result in communication failures between vehicles and with infrastructure elements, affecting safety and efficiency.
5. Regulatory and Legal Challenges:
   - **Liability Issues**: Determining responsibility in the case of a cyberattack or malfunction can be challenging. It raises questions about liability and legal consequences, both for manufacturers and users.
   - **Regulatory Compliance**: Adhering to evolving regulations and standards related to connected vehicles can be challenging for manufacturers, and non-compliance may lead to legal consequences.

Mitigating these risks necessitates a comprehensive approach, involving collaboration among industry stakeholders, implementing robust cybersecurity measures, ensuring privacy safeguards, and continuous regulatory efforts to guarantee the secure and safe integration of connected vehicles into the transportation ecosystem.

1.4 PRIVACY CONCERNS IN IoV

1. Personal Data Collection

   The gathering and dissemination of individual information within the IoV present substantial privacy issues [32]. As vehicles become more connected and data-driven, various sensors and communication systems collect a wealth of information.

   Here are several facets associated with gathering and distributing individual information within the IoV:

- **Location Data**: IoV relies on location data to enable services such as navigation and real-time traffic updates. However, the continuous tracking of a vehicle's location raises privacy concerns.
- **Driver Behavior**: IoV systems can collect information about driving habits such as speed, acceleration, braking patterns, and other variables. While this information can be used for insurance purposes, it also raises concerns about driver security.
- **Infotainment Systems**: Infotainment systems in connected vehicles may store user preferences, such as music choices, contact information, and frequently visited locations.
- **Health and Well-Being**: Some IoV systems may integrate biometric sensors to monitor the driver's health and well-being. This could include heart rate monitoring or other health-related metrics.
- **Call and Messaging Data**: Integrated communication systems may record call logs and messages, potentially infringing on user privacy.

2. Sharing of Personal Data:

   Vehicle-to-Cloud Communication:

   - **Cloud Services**: Data collected by vehicles may be transmitted to cloud servers for storage, analysis, and retrieval. This highlights concerns about data security, both during transmission and retention [33].

3 Vehicle-to-Vehicle Communication (V2V):

   - **Safety Alerts**: Vehicle-to-vehicle (V2V) communication entails exchanging safety-critical details among vehicles, like abrupt stops or roadway dangers.. While this is critical for safety, it involves sharing real-time data that could be privacy-sensitive.

1.4.1 Privacy Challenges in Location Tracking and Vehicle Data

Privacy challenges in location tracking and vehicle data arise from the increasing use of technologies that capture and store information about the movements and activities of individuals [34]. Here are some key privacy challenges in this context (Figure 1.4):

- Data Collection and Retention:
  - **Excessive Data Collection**: Gathering more data than necessary for a specific purpose can lead to unnecessary privacy risks.

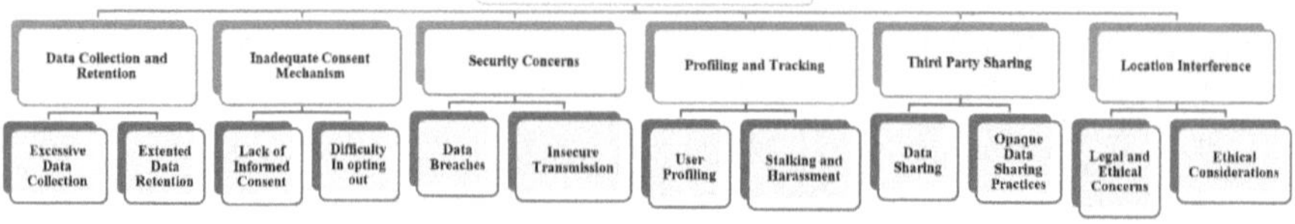

FIGURE 1.4 Privacy challenges in IoV.

    - **Extended Data Retention**: Keeping location and vehicle data for extended periods increases the potential for misuse and unauthorized access.
- Inadequate Consent Mechanisms:
    - **Lack of Informed Consent**: Users may lack a complete understanding of the implications associated with granting permission for location tracking, especially if the information is used beyond its original purpose.
    - **Difficulty in Opting Out**: Users may find it challenging to disable location tracking features or may not be aware of how to do so.
- Security Concerns
    - **Data Breaches**: Keeping confidential location and vehicle information heightens the possibility of data breaches, resulting in unauthorized entry and possible exploitation.
    - **Insecure Transmission**: Data transmitted between devices and servers may be susceptible to interception if proper encryption measures are not in place.
- Profiling and Tracking
    - **User Profiling**: Continuous monitoring and analysis of location data can lead to the creation of detailed user profiles, raising concerns about targeted advertising, discrimination, or surveillance.
- Stalking and Harassment:
    - **Government Surveillance**: Governments may use location and vehicle data for mass surveillance, raising concerns about privacy violations and potential abuse of power.
    - **Lack of Legal Protections**: In some jurisdictions, there may be insufficient legal safeguards to protect individuals from unwarranted government surveillance.
- Third-Party Sharing:
    - **Data Sharing with Third Parties**: Companies may share location and vehicle data with third-party entities, leading to privacy risks if the recipients do not adequately protect the information.
    - **Opaque Data Sharing Practices**: Lack of transparency regarding data-sharing agreements and practices can erode user trust.
- Location Inference:
    - **Legal and Ethical Concerns**: In some regions, there may be a lack of clear regulations governing the collection, use, and storage of location and vehicle data.
    - **Ethical Considerations**: Companies must consider the ethical implications of using location data, balancing business interests with user privacy rights.

To tackle these privacy issues, it's essential to integrate technological remedies, strong legal structures, and ethical reflections, ensuring accountable and clear approaches in overseeing location tracking and managing vehicle data.

1.4.2 Implications of Privacy Breaches in Smart Transportation

Privacy breaches in smart transportation can have far-reaching implications, affecting individuals, organizations, and society at large [35]. Here are some key implications of privacy breaches in the context of smart transportation:

- Individual Privacy Concerns:
  - **Location Tracking**: Unauthorized access to location information can lead to tracking of persons' locations, raising concerns about stalking, harassment, and breach of personal boundaries.
  - **Personal Information Exposure**: Breaches may expose personal details such as home addresses, travel patterns, and frequently visited locations, compromising individuals' privacy.
- Safety and Security Risks:
  - **Physical Security**: If malicious actors gain access to smart transportation systems, they could exploit vulnerabilities to cause accidents or disruptions, posing risks to public safety.
  - **Vehicle Tampering**: Unauthorized access to vehicle data might enable attackers to tamper with critical systems, affecting the safety and performance of vehicles.
- Identity Theft and Fraud:
  - **Data Misuse**: Stolen personal and financial information from smart transportation systems can be used for identity theft, fraud, or other malicious activities.
  - **Payment and Transaction Security**: Breaches in payment systems integrated into smart transportation solutions may lead to unauthorized transactions and financial losses.
- Public Trust Erosion:
  - **Loss of Confidence**: High-profile privacy breaches can erode public trust in smart transportation technologies, making individuals reluctant to adopt or use these systems.
  - **Negative Perception**: Privacy concerns can lead to a negative perception of smart transportation initiatives, hindering their widespread acceptance and adoption.
- Legal and Regulatory Consequences:
  - **Legal Liability**: Organizations responsible for privacy breaches may face legal action, lawsuits, and regulatory penalties for failing to protect user data.
  - **Regulatory Scrutiny**: Incidents of privacy breaches could prompt increased regulatory scrutiny and the introduction of new, stricter regulations for smart transportation systems.
- Business Reputational Damage:
  - **Brand Image**: Privacy breaches can tarnish the reputation of companies involved in smart transportation, affecting their brand image and market standing.

  - **Customer Loyalty**: Concerns about privacy may lead to a loss of customer loyalty, as individuals may seek alternative transportation options that they perceive as more secure.
- Data Manipulation and Misuse:
  - **False Information**: Breaches may lead to the manipulation of transportation data, potentially causing misinformation about traffic conditions, vehicle locations, or routes.
  - **Operational Disruption**: Data manipulation could disrupt the efficient operation of smart transportation systems, leading to logistical challenges and service interruptions.
- Social and Economic Impact:
  - **Productivity Losses**: Disruptions caused by privacy breaches can result in economic losses and impact the productivity of smart transportation systems.
  - **Unequal Access**: Privacy concerns may disproportionately affect vulnerable populations, leading to disparities in access to and benefits from smart transportation technologies.

To address these concerns, it's essential for entities and policymakers to emphasize strong cybersecurity protocols, incorporate privacy-centric design principles, and set forth explicit regulations ensuring user privacy within the intelligent transportation environment. Moreover, educating users and initiating awareness campaigns can assist individuals in comprehending the potential hazards and precautionary steps related to intelligent transportation technologies.

1.4.3 The Role of Encryption in Protecting Personal Information in Connected Transportation

Encryption is critical for protecting personal data across networked transportation systems, ensuring the confidentiality and accuracy of data while it is sent and stored within these platforms [36]. Here are some important details illustrating the significance of encryption in securing personal information in the field of connected transportation:

- Secure Communication:
  - **Data in Transit**: Encryption ensures that data exchanged between connected vehicles, infrastructure, and backend servers are secured during transmission. This prevents unauthorized interception and eavesdropping by malicious actors.
- Data Integrity:
  - **Tamper Resistance**: Encryption aids in preserving the integrity of data by rendering it challenging for unauthorized parties to alter or tamper with information during transit. This ensures that the data received are the same as the data sent.

- Protecting User Identities:
  - **User Authentication**: Encryption is used in authentication processes to verify the identities of users, vehicles, and devices, preventing unauthorized access to sensitive information and systems.
  - **Credentials Protection**: Encrypted storage of user credentials, such as login details and access tokens, helps prevent unauthorized access even if physical or digital breaches occur.
- Securing V2I and V2V Communication:
  - **V2I Communication**: Encryption ensures the integrity and confidentiality of vehicle and infrastructure communication, including traffic lights, road signs, and other connected components.
  - **V2V Communication**: Encrypting communications between vehicles helps protect sensitive information exchanged for safety and coordination purposes, such as collision warnings or traffic updates.
- Protection Against Eavesdropping:
  - **Man-in-the-Middle Attacks**: Encryption guards against attacks known as "man-in-the-middle," in which an uninvited party intercepts and perhaps modifies communication between connected devices or systems.
- Securing Location Data:
  - **Geolocation Privacy**: Encryption helps safeguard people's privacy by securing location data and avoiding unapproved parties from tracking the movements of connected vehicles and their occupants.
- Data-at-Rest Protection:
  - **Storage Encryption**: Personal information stored in databases or onboard systems is protected through encryption, reducing the risk of data breaches in case of actual robbery or unapproved admittance to storage devices.
- Compliance with Privacy Regulations:
  - **Legal Requirements**: Numerous security guidelines, including the General Data Protection Regulation (GDPR) and the California Consumer Privacy Act (CCPA), mandate the use of encryption as part of comprehensive data protection measures. Adhering to these regulations is essential for legal compliance.
- Minimizing Insider Threats:
  - **Internal Access Controls**: Encryption, when combined with strong access controls, helps mitigate risks posed by insider threats, limiting the ability of authorized personnel to misuse or access sensitive personal information without proper authorization.
- Building User Trust:
  - **User Confidence**: The implementation of robust encryption measures fosters user trust in connected transportation systems. Users are more likely to adopt and utilize these systems if they are confident that their personal information is securely handled.

While encryption is a powerful tool for enhancing the security of connected transportation systems, it should be complemented by other cybersecurity measures, including regular security audits, software updates, and a privacy-by-design approach to ensure comprehensive protection against evolving threats [37].

1.5 SECURITY MEASURES IN IoV

1.5.1 Encryption and Authentication in IoV

In the realm of the IoV, marked by the interconnectivity of vehicles, infrastructure, and other entities, the essential roles of encryption and authentication take center stage. They ensure the security and privacy of both data and communications [38,39]. Here's an overview of how encryption and authentication are applied in IoV:

Encryption in IoV:

- Data in Transit Protection:
  - **V2V Communication**: Communication between vehicles entails the exchange of safety-critical information. Encryption ensures that this communication is secure, preventing eavesdropping and tampering with the transmitted data.
  - **V2I Communication**: Communication among vehicles and infrastructure(e.g., traffic lights, road signs, and control centers) is secured through encryption, protecting the integrity and confidentiality of the exchanged data.
- Secure Firmware and Software Updates:
  - **Over-the-Air (OTA) Updates**: Encryption is essential for securing firmware and software updates sent to vehicles over the air, preventing unauthorized access and tampering during the update process.
- Protection of Location Data:
  - **Geolocation Privacy**: Encryption helps safeguard the privacy of location data generated by vehicles, Ensuring that authorized entities are the sole entities with access to this sensitive information.
- Secure Mobile Applications:
  - **Mobile App Communication**: IoV often involves mobile applications that interact with vehicles. Encryption secures the communication between the mobile app and the vehicle, protecting user data and commands.
- Securing Telematics Data:
  - **Telematics Systems**: Telematics systems, which collect and transmit vehicle-related data, utilize encryption to uphold the secrecy and integrity of this information.
- Authentication Tokens:
  - **Tokenization**: Encryption is used to tokenize authentication tokens exchanged between vehicles and backend servers, ensuring that sensitive information, such as access tokens, remains secure.
- Preventing Replay Attacks:
  - **Nonce and Timestamps**: Encryption, combined with nonces (random numbers used only once) and timestamps, helps prevent replay attacks, where intercepted messages are replayed to gain unauthorized access.
- Secure Cloud Connectivity:
  - **Cloud Services**: Many IoV services involve cloud connectivity. Encryption ensures the secure transmission of data between vehicles and cloud servers, protecting against unauthorized access.

Authentication in IoV:

- Vehicle Identity Verification:
  - **V2V and V2I Authentication**: Vehicles need to authenticate themselves to each other and to infrastructure components. Authentication ensures that only trusted entities can participate in IoV communication.
- User Identity Verification:
  - **Driver Authentication**: Authentication mechanisms verify the identity of drivers interacting with in-vehicle systems or mobile applications, keeping out unwanted access.
- Secure Access Control:
- **Role-Based Access**: Authentication links with access control systems, guaranteeing that solely designated individuals or systems can reach particular functions or information within the IoV environment.
- Credential Protection:
  - **Secure Storage**: Authentication credentials, such as private keys and passwords, are stored securely using encryption, minimizing the risk of unauthorized access even if physical or digital breaches occur.
- Multi-Factor Authentication (MFA):
  - **Enhanced Security**: Security is bolstered through multi-factor authentication, which requires the verification of multiple elements, such as a password and a biometric scan. This approach provides added protection against unauthorized access.
- Secure Communication Channels:
  - **Secure Channels**: Authentication is utilized to lay out secure communication channels between vehicles, infrastructure, and backend servers, ensuring that only authenticated entities can exchange sensitive information.
- Digital Signatures:
  - **Message Integrity**: Digital signatures, often used in conjunction with authentication, authenticate the origin and integrity of messages exchanged within the IoV ecosystem.

By incorporating strong encryption and authentication protocols, IoV platforms can reduce the chances of illicit entry, data breaches, and manipulation, creating a reliable and secure environment for interconnected vehicles and associated services. As the IoV environment progresses, maintaining alertness and embracing updated security methodologies become essential to tackle new vulnerabilities.

1.5.2 Intrusion Detection and Prevention Systems

Intrusion detection systems (IDS) and intrusion prevention systems (IPS) stand as crucial elements within cybersecurity infrastructure, designed to detect and respond to security incidents and attacks [40]. These systems play a pivotal function in protecting networks, systems, and applications against unauthorized access, malicious activities, and potential threats.

- Intrusion Detection Systems:
  - Functionality:
    - IDS continuously observes activities within networks and systems in real-time, searching for patterns or behaviors that could signal a security incident or potential intrusion.
  - Detection Methods:
    - **Signature-Based Detection**: It contrasts observed patterns with a repository of recognized attack signatures.
    - **Anomaly-Based Detection**: It creates a standard for typical activity and triggers alerts upon detecting deviations from this established baseline.
    - **Behavior-Based Detection**: It scrutinizes user and system behavior to identify suspicious activities.
  - Deployment:
    - **Network-Based IDS (NIDS)**: It oversees network traffic and scrutinizes packets for indications of malicious activity.
    - **Host-Based IDS (HIDS)**: It observes activities on individual hosts or devices, looking for signs of compromise.
  - Response:
    - **Alerts**: It generates alerts or notifications when potential intrusions or security incidents are detected.
    - **Logs**: It records information about detected incidents for further analysis and investigation.
- Intrusion Prevention Systems:
  - **Functionality**: IPS builds upon IDS by not only detecting but also actively preventing or blocking identified threats in real time.
  - Prevention Methods:
    - **Packet Filtering**: Blocks or allows network packets based on predefined rules.
    - **Signature-Based Prevention**: Blocks known malicious signatures before they can reach their targets.
    - **Behavior-Based Prevention**: Takes proactive measures to stop activities that exhibit malicious behavior.
  - Deployment:
    - **Network-Based IPS (NIPS)**: Sits on the network perimeter and can block malicious traffic before it reaches its destination.
    - **Host-Based IPS (HIPS)**: Operates on individual devices and can prevent malicious activities at the host level.
  - Response:
    - **Automated Blocking**: Automatically blocks or mitigates malicious activities based on predefined rules.
    - **Alerts and Logs**: Like IDS, IPS generates alerts and logs for monitoring and analysis.

- Integration with Security Ecosystem:
  - **SIEM Integration**: IDS and IPS are often integrated with security information and event management (SIEM) systems for centralized monitoring and analysis.
  - **Collaboration with Firewalls**: Coordinated efforts with firewalls and other security measures enhance overall network security.

Deploying a combination of intrusion detection and prevention measures is crucial for maintaining a robust cybersecurity posture and safeguarding against a wide range of cyber threats. Regular updates, proper configuration, and a comprehensive understanding of the network environment are essential for the effectiveness of IDS and IPS solutions [41].

1.5.3 Securing V2X Communication

Securing V2X communication is crucial to protect the confidentiality, integrity, and reliability of the information swapped between vehicles and their surroundings. V2X communication includes V2V, V2I, V2P, as well as additional channels of communication that contribute to the connected and autonomous vehicle ecosystem [42,43]. Here are key considerations for securing V2X communication:

- Encryption:
  - **End-to-End Encryption**: Utilize end-to-end encryption to safeguard the secrecy of data exchanged among vehicles, infrastructure, and various entities. This hinders illicit access to vital details like location specifics, personal identification, and crucial safety messages.
- Authentication:
  - **Vehicle Authentication**: Convey vigorous validation systems to ensure that main approved and genuine vehicles can participate in V2X communication. This may involve the use of digital certificates, secure key exchange protocols, and multi-factor authentication.
  - **Infrastructure Authentication**: Authenticate communication with infrastructure components to forestall unapproved access and malicious manipulation of data. This includes verifying the legitimacy of traffic signals, road signs, and other infrastructure elements.
- Secure Key Management:
  - **Key Distribution**: Institute secure protocols for the distribution and administration of cryptographic keys. Effective key management is vital for upholding the security of encrypted communications.
  - **Key Rotation**: Periodically rotate encryption keys to reduce the risk associated with long-term key compromises. Regularly updating keys helps maintain the security of the communication infrastructure.
- Secure Protocols:
  - **Use Standardized Protocols**: Implement standardized and widely accepted communication protocols, such as DSRC or C-V2X. These

protocols typically incorporate inherent security features and have undergone rigorous testing processes.

- **Transport Layer Security (TLS)**: When using wireless communication, implement TLS to secure data in transit. TLS ensures the integrity and confidentiality of communication channels.

- Network Security:
  - **Firewalls and Intrusion Detection/Prevention Systems**: Deploy security measures for networks, including firewalls and systems for intrusion detection and prevention, to monitor and command incoming and outgoing traffic. These systems can detect and mitigate potential security threats.
- Privacy Protection:
  - **Pseudonymization**: Implement techniques like pseudonymization to protect the privacy of individuals by using temporary identifiers instead of permanent ones.
  - **Minimal Data Collection**: Limit the collection of personal or sensitive data to the minimum required for V2X functionality. Incorporate privacy-by-design principles into the development of V2X systems.

Securing V2X communication requires a holistic and collaborative approach involving manufacturers, infrastructure providers, policymakers, and cybersecurity experts. Regular security assessments, updates, and collaboration within the automotive and transportation industries are necessary to keep up with new threats and vulnerabilities.

1.6 PRIVACY PROTECTION STRATEGIES

1.6.1 Anonymization and Pseudonymization Techniques

Anonymization and pseudonymization are methods employed to preserve individuals' confidentiality by either obscuring or substituting personally identifiable information (PII) within data collections. These strategies find application in sectors like healthcare, finance, and research to adhere to privacy norms and minimize the potential exposure of confidential details [44].

Anonymization: This technique alters or eliminates PII to ensure individuals cannot be discerned from the dataset. The objective is to obscure individual identities entirely. Various anonymization methods encompass:

- **Data Aggregation**: Merging data to eliminate specific individual elements.
- **Generalization**: Simplifying data details while preserving its foundational structure.
- **Randomization**: Injecting variability into data, like distorting numerical values.
- **Data Masking/Tokenization**: Substituting sensitive data with generic placeholders or tokens.
- **Data Swapping**: Shuffling data values across records to disconnect them from specific individuals.

However, it's crucial to recognize that achieving full anonymization presents challenges, necessitating careful evaluation of potential re-identification risks.

Pseudonymization:
Pseudonymization revolves around substituting or encoding personally identifiable details with pseudonyms or codes. Unlike full anonymization, re-identifying individuals remains feasible using separate ancillary data, like a decryption key. Pseudonymized [45] data can often revert to its original form when coupled with the initial data. Key methods of pseudonymization include:

- **Tokenization**: Exchanging confidential data with distinct tokens or identifiers.
- **Hashing**: Employing irreversible processes to convert data into a consistent-length string (hash).
- **Encryption**: Converting data via an algorithm and specific key, rendering it unreadable without its corresponding decryption mechanism.
- **Dynamic Masking**: Revealing only segments of sensitive data, obscuring the remainder using placeholders like asterisks.

Pseudonymization is apt for situations requiring certain data connections while prioritizing individual privacy through complicating direct data-person associations.

Often, blending anonymization and pseudonymization strategies offers heightened privacy, facilitating beneficial data analysis and sharing. The contextual privacy demands and governing regulations of the pertinent sector should guide the choice and implementation of these methods.

Techniques for protecting user identities
Ensuring user identity protection is paramount for maintaining privacy and preventing unauthorized breaches into confidential data. To bolster this protection, a multifaceted approach is essential [44,43]. This includes advanced authentication methods like multi-factor and biometric authentication, alongside robust encryption techniques such as end-to-end encryption and tokenization. Access controls like role-based access and the least privilege principle further limit unauthorized access. Additionally, strategies like anonymization, pseudonymization, and strict password policies ensure data confidentiality and user anonymity. Continuous user education, monitoring, adherence to secure communication protocols, and a commitment to privacy-by-design principles, along with regulatory compliance, collectively form a comprehensive security framework. This integrated approach significantly fortifies defenses against identity threats while upholding user privacy.

1.6.1.1 Challenges and Limitations of Anonymization

Anonymization is vital for safeguarding privacy but presents inherent challenges. Even when data are anonymized, risks of re-identification persist when merged with external datasets or inadequately randomized, potentially revealing individuals. Additionally, anonymization can degrade data quality, strip contextual relevance, and become less effective over time due to evolving datasets or emerging re-identification techniques. Furthermore, highly anonymized data may not support advanced analytics or satisfy regulatory definitions across jurisdictions. Moreover, there's a risk of users misinterpreting the security of anonymized data, and ethical concerns remain, especially with sensitive information. Given these complexities, organizations must holistically approach data privacy, combining anonymization methods [43] with stringent security protocols, consistent risk evaluations, and adherence to evolving privacy [46] regulations, all while balancing privacy and data utility.

1.6.2 Consent Management in IoV

Consent management regarding the IoV refers to the systematic approach of obtaining, maintaining, and managing user permissions for data collection, processing, and sharing within the vehicular network. With the extensive volumes of data generated and exchanged among vehicles, infrastructure, and other connected devices in IoV ecosystems, ensuring that participants have control over their data is paramount.

Here are some key aspects and considerations of consent management in IoV:

1. **Explicit Consent**: Before collecting or processing personal data from vehicles or drivers, explicit consent should be obtained. This ensures that individuals are provided with information about the collected data, its intended purpose, and the entities with whom it may be shared.
2. **Granular Consent Options**: Individuals should have the choice to grant consent flexibly for specific data types or purposes rather than giving broad, blanket consent. For instance, a driver might consent to share location data for traffic management but may opt out of sharing data related to personal habits or preferences.
3. **Revocable Consent**: Users should possess the right to retract their consent at any given moment and establish mechanisms that enable individuals to effortlessly withdraw previously granted consent, halting further data processing or sharing.
4. **Transparent Policies**: Organizations operating IoV systems should maintain transparent data usage policies that clearly outline how consent is obtained; how data are used, processed, and stored; and the rights of individuals regarding their data.
5. **Consent Management Platforms**: Utilizing advanced technologies and platforms can help automate and streamline the consent management process. These platforms can track user preferences, manage consent settings, and ensure compliance with regulatory requirements.
6. **Compliance with Regulations**: Organizations are required to agree with material information insurance and security guidelines, such as the General Data Protection Regulation (GDPR) in Europe or other applicable regional data protection laws. These regulations often have strict requirements concerning consent management, data protection, and user rights.
7. **Education and Awareness**: Ongoing user education and awareness initiatives can assist individuals in comprehending the significance of consent in

IoV systems. By empowering users with knowledge about their data rights and how their information is used, organizations can foster trust and collaboration within the IoV ecosystem.

In summary, consent management in IoV is essential to respect user privacy, ensure data protection, and adhere to regulatory requirements. By implementing robust consent management practices, organizations can build trust with users, promote responsible data usage, and facilitate the secure and ethical development of IoV technologies.

1.7 FUTURE VISION AND CHALLENGES

The evolution of the IoV brings forth opportunities and challenges, with a particular focus on security and protection. Several emerging trends are shaping the future of IoV security and privacy:

- **Zero Trust Architecture (ZTA)**: As IoV ecosystems grow more complex, the adoption of zero trust architecture is gaining momentum. This approach emphasizes strict identity verification and continuous authentication, making certain that only duly authorized and recognized entities can access resources within the IoV environment, thereby reducing potential vulnerabilities.
- **Edge Computing**: As connected vehicles proliferate, generating vast amounts of data, edge computing is emerging as a pivotal trend. Edge computing [47] minimizes latency and boosts real-time decision-making capabilities by handling data in proximity to its origin. At the same time, it reduces the risk linked to transmitting sensitive data across networks.
- **AI and ML**: Utilizing the capabilities of AI and ML algorithms [43,48] can bolster IoV security by identifying anomalous patterns, predicting potential cyber threats, and enabling autonomous responses. These technologies can also enhance privacy by implementing advanced encryption techniques and data anonymization strategies.
- **Blockchain Technology**: The decentralized and immutable nature of blockchain holds the potential for improving IoV security [38] and privacy. By creating tamper-proof records of transactions and interactions within the IoV ecosystem, blockchain can mitigate risks associated with data tampering, unauthorized access, and fraudulent activities [49,50,51].
- **Enhanced Data Encryption**: As the volume of data transmitted across IoV networks escalates, implementing robust encryption algorithms and protocols is imperative. Advanced encryption techniques, including homomorphic encryption and quantum-resistant cryptography, are emerging as essential tools to protect sensitive information and ensure data privacy [52].
- **Structures for Regulation**: Recognizing the criticality of security and privacy in IoV implementations, regulatory bodies worldwide are actively developing comprehensive frameworks and standards. These regulations

aim to establish baseline security requirements, promote industry-wide compliance, and foster trust among stakeholders.

- **Privacy-Preserving Technologies**: To address growing concerns surrounding data privacy, innovative technologies that are becoming more popular integrate secure multi-party computation, federated learning, and differential privacy [45,53]. These privacy-preserving techniques enable data analysis while preserving individual user privacy and confidentiality.
- **Threat Intelligence Sharing**: Collaborative efforts among automotive manufacturers, cybersecurity firms, and government agencies are fostering the sharing of threat intelligence and best practices. This collaborative approach enables the timely identification of emerging threats, facilitates knowledge exchange, and enhances the IoV ecosystem's overall security posture.

As IoV technologies continue to advance and proliferate, addressing evolving security and privacy challenges remains paramount. By embracing these emerging trends and adopting a proactive strategy for cybersecurity and privacy, stakeholders can navigate the complexities of the IoV landscape while guaranteeing the security and privacy of connected vehicles and their occupants.

1.7.1 Security and Privacy Measures

Future research in the IoV should prioritize enhancing security and privacy measures [54], refining standardization for seamless interoperability, bolstering resilience against cyber threats, and harnessing edge computing for real-time data processing. Additionally, advancements in machine learning, AI techniques, and human-machine interaction are essential for optimizing traffic management, user experience, and autonomous vehicle functionality. Comprehensive regulatory frameworks, infrastructure optimization, and sustainability considerations are pivotal to address legal, environmental, and energy efficiency challenges. Collaborative strategies, user acceptance studies, robustness testing, and ethical evaluations further underscore the multifaceted approach required to ensure the responsible and efficient development of IoV technologies, emphasizing interdisciplinary collaboration and stakeholder engagement.

1.7.2 Ethical Implications of IoV Technology

The IoV introduces a myriad of ethical considerations encompassing privacy [46,55], security, autonomous decision-making, equity, user autonomy, transparency, economic impact, environmental sustainability, social trust, and long-term consequences. Key concerns include safeguarding user privacy amid extensive data collection, addressing cybersecurity vulnerabilities and ensuring data integrity, grappling with ethical dilemmas in autonomous decision-making and liability attribution, mitigating potential digital divides and affordability challenges, preserving user control over automated systems, ensuring system transparency and accountability, addressing job displacement concerns, balancing technological advancements with environmental stewardship, fostering public trust, and anticipating unintended societal impacts. Tackling these diverse ethical considerations requires interdisciplinary collaboration among technologists, ethicists, policymakers, and the public. Together, they can formulate ethical frameworks prioritizing individual and societal well-being, guiding the responsible development and deployment of innovations in the IoV.

1.7.3 Balancing Security and Privacy with Societal Concerns

Balancing security and privacy concerns with societal implications in the IoV smart transportation system is a difficult task that needs careful consideration of many different aspects. On the one hand, ensuring strong security protocols [46] is paramount to protect against cyber threats, unauthorized entry, and data breaches, thereby safeguarding the integrity, reliability, and safety of the transportation ecosystem. Implementing advanced encryption protocols, secure communication channels, and robust authentication mechanisms can bolster security while preserving data confidentiality and system integrity.

Concurrently, addressing privacy concerns is essential to protect individuals' rights, autonomy, and personal information. Establishing stringent data protection protocols, anonymization techniques, user consent frameworks, and transparency measures can mitigate privacy risks and foster user trust. However, finding the ideal balance between privacy and security necessitates navigating potential trade-offs, ethical dilemmas, and societal implications.

Moreover, considering societal concerns [55,56] involves addressing broader issues such as equity, accessibility, economic impact, public trust, ethical considerations, and regulatory compliance. Ensuring equitable access to IoV technologies, promoting affordability, addressing digital divides, preserving user autonomy, fostering public acceptance, and adhering to regulatory frameworks are essential to align IoV developments with societal values and norms.

Therefore, achieving a harmonious balance requires a holistic approach that integrates technological innovations, ethical considerations, policy frameworks, stakeholder engagement, and public participation. Cooperative actions between industrial participants, policymakers, ethicists, technologists, and the public are vital to navigate these complexities effectively, prioritize societal well-being, and foster a responsible and sustainable IoV smart transportation system that harmoniously integrates security, privacy, and societal concerns.

1.8 THE FRAMEWORK FOR BLOCKCHAIN-INTEGRATED INTERNET OF VEHICLE SYSTEM

1.8.1 Decentralized Ledger Technology: Blockchain

Blockchain technology [57] is a network of peers that stores transaction data in a secure ledger using consensus and cryptography techniques. Every block in the chain carries the transaction data, a timestamp, and the encryption key of the previous block. The use of the Merkle tree algorithm ensures tamper-resistant content within the block. This decentralized ledger, associated with blockchain, allows transparent

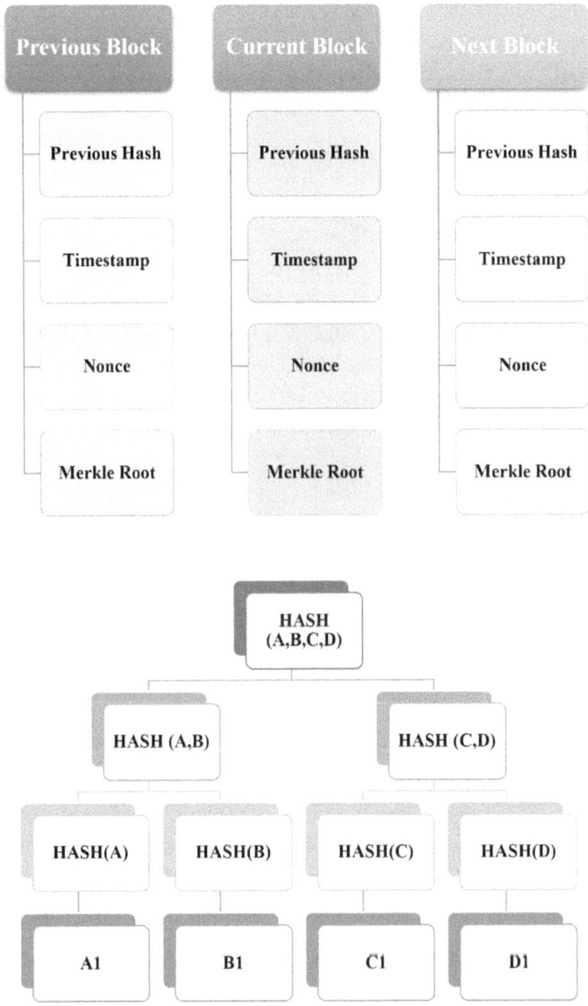

FIGURE 1.5 (a and b) Blockchain data layer.

and permanent recording of transactions. Its prominent application lies in digital currencies like Bitcoin, where it serves as a public ledger for tracking transactions and account balances (Figure 1.5).

1.8.1.1 Integration of Blockchain in Autonomous Vehicles (AVs)

1. **Self-driving Technology**: Blockchain stores information gathered via vehicle sensors enhancing transparency and reducing data theft risks. It enables monitoring and sharing of vehicle safety information and usage patterns among relevant parties.

2. **Telematics**: It utilizes blockchain for secure data sharing and communication between various self-driving cars and organizations, such as local governments. It enhances transparency, efficiency, and security in data sharing across AVs.

1.8.1.1.1 Considerations in Blockchain Application for AVs

Vehicle Safety: Sensors identify instances of speeding, alert the driver, or assume control of the vehicle to prevent negligence, serving as the foundation for determining insurance premiums. Driving behaviors are evaluated using sensor data that are kept on the blockchain, improving security [58].

Accident Management: Blockchain aids [39] in automatic positioning and emergency assistance by providing relevant information to rescue organizations. It helps determine the exact location and severity of accidents, expediting rescue efforts.

It combines wireless communication technologies and global satellite positioning for all-encompassing vehicle tracking. Monitors emergency alerts, overloading, driving when fatigued, and routes.

1.8.2 Challenges in Blockchain Applications for AVs

- **Anonymity Challenges:** Challenges arise in verifying the accuracy of data sources due to the anonymity inherent in blockchain data. Lack of identification for vehicles poses challenges in managing a comprehensive system.
- **Security, Scale, and Execution Efficiency:** Struggling to strike a balance between security, scalability, and operational efficiency, the popularity and technical significance of blockchain technology in AV applications present challenges.
- **Decentralized Identifier (DID) and Smart Contract Issues:** DID technology for vehicle identification is not widely used. Blockchain networks's performance in the extremely mobility atmosphere of autonomous vehicle networks is a source of concern. Smart contract specifications lack widespread availability, posing challenges in regulating sensor data and vehicular networking technologies.
- **Big Data Challenges:** The challenges include managing the huge volume of data produced by multiple apps in the big data era and addressing the high computing requirements of blockchain technology in the AV infrastructure.

As blockchain technology [54,59] continues to evolve, addressing these challenges is crucial for realizing its full potential in improving the safety, transparency, and efficiency of autonomous vehicles. Continuous research and development are essential to conquer these hurdles and optimize the integration of blockchain in the AV ecosystem.

1.8.3 BC-Based Application for IoV

Decentralization stands out as a pivotal feature of blockchain technology [45], offering advantages such as transparency, affordability, immutability, and enhanced confidentiality. This makes blockchain a viable and cost-efficient solution for fortifying the IoV system, ensuring its reliability and privacy. Specifically, four significant applications have emerged for blockchain-integrated IoV: security, credibility, credit-based incentives, and privacy preservation. In the realm of safety, trustworthiness, and confidentiality, applications are clearly delineated, based on credit incentives that motivate. Entities within the IoV [54,60] collaborate by sharing their computational and networking resources, earning credits for their valuable contributions. IoV connects numerous vehicles for data sharing on collisions, traffic updates, weather conditions, and infotainment, presenting challenges in managing vast data

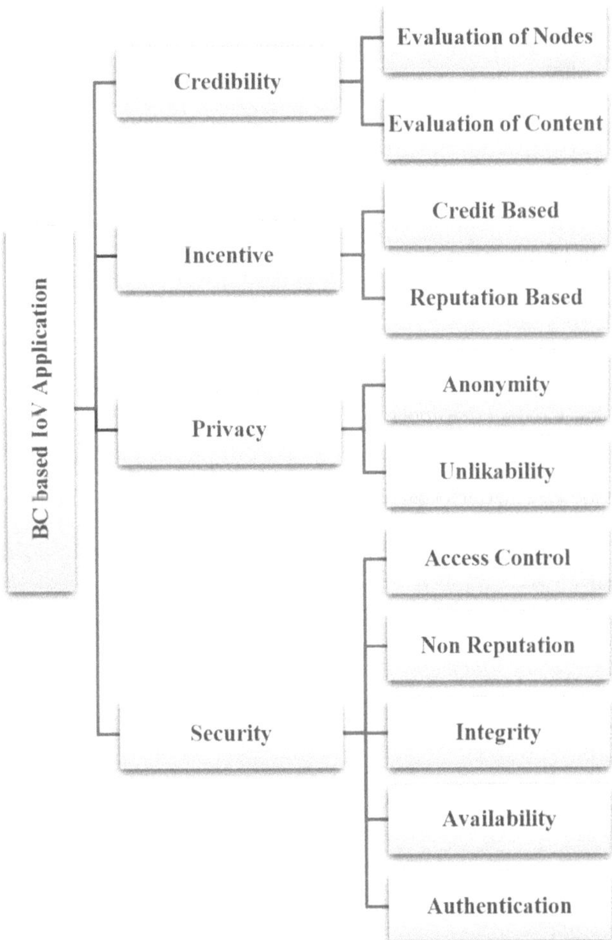

FIGURE 1.6 BC-based IoV.

volumes without compromising scalability or security. In this context, blockchain addresses essential IoV security requirements like data provenance, transparency, and resilience without central authority intervention. By doing so, blockchain fosters trust among vehicles in intricate environments, ensuring data integrity, resilience against cyberattacks, and user anonymity. Additionally, the research underscores blockchain's efficacy in safeguarding IoV security and privacy, employing techniques like k-anonymity to protect vehicle privacy by making individual vehicles indistinguishable within groups (Figure 1.6).

1.9 CONCLUSION

1.9.1 Summary Key Findings

The integration of AI into IoV security presents transformative possibilities for vehicular communication and the automotive sector, although it introduces notable challenges such as privacy risks, infrastructure complexities, standardization needs, and scalability issues. Future research endeavors should prioritize advanced AI modeling, blockchain incorporation, interdisciplinary collaboration, and privacy-centric methodologies, necessitating a multifaceted approach involving AI specialists, IoV experts, policymakers, and users. Despite advancements, past IoV incidents underscore the criticality of addressing cybersecurity vulnerabilities through proactive security measures, industry collaboration, user education, privacy-enhancing technologies, regulatory compliance, and continuous vigilance to navigate the intricate balance between connectivity and security effectively.

REFERENCES

1. F.-Y. Wang, Y. Lin, P. A. Ioannou, L. Vlacic, X. Liu, A. Eskandarian, Y. Lv, X. Na, D. Cebon, J. Ma, L. Li, and C. Olaverri-Monrea, "Transportation 5.0: the DAO to safe, secure, and sustainable intelligent transportation systems" *IEEE Transactions on Intelligent Transportation Systems*, vol. 24, no. 10, pp. 10268–10278, Oct. 2023.
2. H. Zhou, W. Xu, J. Chen, and W. Wang, "Evolutionary V2X technologies toward the internet of vehicles: challenges and opportunities," *Proceedings of the IEEE*, vol. 108, no. 2, pp. 308–323, Feb. 2020.
3. S. Djahel, R. Doolan, G.-M. Muntean, and J. Murphy, "A communications-oriented perspective on traffic management systems for smart cities: challenges and innovative approaches," *IEEE Communications Surveys & Tutorials*, vol. 17, no. 1, pp. 125–151, 1st Quart., 2015.
4. Md. J. N. Mahi, S. Chaki, S. Ahmed, M. Biswas, M. Shamim Kaiser, M. S. Islam, M. Sookhak, A. Barros, and Md. Whaiduzzaman, "A review on VANET research: perspective of recent emerging technologies," *IEEE Access*, vol. 10, pp. 65760–65783, 2022.
5. L. Chettri, and R. Bera, "A comprehensive survey on Internet of Things (IoT) toward 5G wireless systems," *IEEE Internet of Things Journal*, vol. 7, no. 1, pp. 16–32, Jan. 2020.
6. J. Contreras-Castillo, S. Zeadally, and J. A. Guerrero-Ibañez, "Internet of vehicles: architecture, protocols, and security," *IEEE Internet of Things Journal.*, vol. 5, no. 5, pp. 3701–3709, Oct. 2017.
7. Y. Fangchun, W. Shangguang, L. Jinglin, L. Zhihan, and S. Qibo, "An overview of internet of vehicles," *China Communications*, vol. 11, no. 10, pp. 1–15, Oct. 2014.

8. R. H. Weber, "Internet of things – new security and privacy challenges," *Journal of Computational Law and Security Review*, vol. 26, no. 1, pp. 23–30, 2010.
9. M. Riley, K. Akkaya, and K. Fong, "A survey of authentication schemes for vehicular ad hoc networks," *Security and Communication Networks*, vol. 4, no. 10, pp. 1137–1152, 2011.
10. A. M. Vegni, and V. Loscri, "A survey on vehicular social networks," *IEEE Communications Surveys & Tutorials*, vol. 17, no. 4, pp. 2397–2419, 4th Quart., 2015.
11. K. M. Alam, M. Saini, and A. E. Saddik, "Toward social internet of vehicles: concept, architecture, and applications," *IEEE Access*, vol. 3, pp. 343–357, 2015.
12. F. Cunha, L. Villas, A. Boukerche, G. Maia, A. Viana, R. A. Mini, and A. A. Loureiro, "Data communication in VANETs: protocols, applications and challenges," *Ad Hoc Networks*, vol. 44, pp. 90–103, Jul. 2016
13. A. Al-Fuqaha, M. Guizani, M. Mohammadi, M. Aledhari, and M. Ayyash, "Internet of things: a survey on enabling technologies, protocols, and applications," *IEEE Communications Surveys & Tutorials.*, vol. 17, no. 4, pp. 2347–2376, 1st Quart., 2015.
14. M. R. Palattella, Dohler, A. Grieco, G. Rizzo, J. Torsner, T. Engel, and L. Ladid, "Internet of things in the 5G era: enablers, architecture, and business models," *IEEE Journal on Selected Areas in Communications*, vol. 34, no. 3, pp. 510–527, Mar. 2016.
15. W. Anwar, N. Franchi, and G. Fettweis, "Physical layer evaluation of V2X communications technologies: 5G NR-V2X, LTEV2X, IEEE 802.11bd, and IEEE 802.11p," in *Proceedings of the IEEE Vehicular Technology Conference (VTC)*, Honolulu, HI, USA, Sep. 2019, pp. 1–7.
16. S. Gyawali, S. Xu, Y. Qian, and R. Q. Hu, "Challenges and solutions for cellular based V2X communications," *IEEE Communications Surveys & Tutorials.*, vol. 23, no. 1, pp. 222–255, 1st Quart., 2021.
17. M. Gerlach, A. Festag, T. Leinmuller, G. Goldacker, and C. Harsch, "Security architecture for vehicular communication," in *Presented at the WIT*, Hamburg, Germany, 2006.
18. J. T. Isaac, S. Zeadally, and J. S. Camara, "Security attacks and solutions for vehicular ad hoc networks," *IET Communication Journals*, vol. 4, no. 7, pp. 894–903, Apr. 2010.
19. L. Gafencu, and L. Scripcariu, "Security issues in the internet of vehicles," in *Proceedings of International Conference of Communications (COMM)*, Bucharest, Romania, Jun. 2018, pp. 441–446.
20. C.-M. Chen, B. Xiang, Y. Liu, and K.-H. Wang, "A secure authentication protocol for Internet of vehicles," *IEEE Access*, vol. 7, pp. 12047–12057, 2019
21. C. Lin, D. He, X. Huang, K.-K.-R. Choo, and A. V. Vasilakos, "BSeIn: a blockchain-based secure mutual authentication with fine-grained access control system for industry 4.0," *Journal of Network and Computer Applications*, vol. 116, pp. 42–52, Aug. 2018.
22. M. Sookhak, H. Tang, Y. He, and F. R. Yu, "Security and privacy of smart cities: a survey, research issues and challenges," *IEEE Communications Surveys & Tutorials*, vol. 21, no. 2, pp. 1718–1743, 2nd Quart., 2019.
23. A. Slama, I. Lengliz, and A. Belghith, "TCSR: an AIMD trust-based protocol for secure routing in VANET," in *Proceedings of International Conference on Smart Communincations and Netwworking (SmartNets)*, Yasmine Hammamet, Tunisia, Nov. 2018, pp. 1–8.
24. Y. Liu, Y. Wang, and G. Chang, "Efficient privacy-preserving dual authentication and key agreement scheme for secure V2V communications in an IoV paradigm," *IEEE Transactions on Intelligent Transportation Systems*, vol. 18, no. 10, pp. 2740–2749, Oct. 2017.
25. P. Kumar Sharma, B. Kumar, and S. S. Tyagi, "Security enhancement through flow-based centralized control in SDN," in *IEEE 13th International Conference on Cloud Computing, Data Science & Engineering (Confluence)*, Noida, pp. 40–45, 2023.

26. P. K. Sharma, B. Kumar, and S. S. Tyagi, "STADS: security threats assessment and diagnostic system in software defined networking (SDN)," in *IEEE International Conference on Machine Learning, Big Data, Cloud and Parallel Computing (COM-IT-CON)*, Faridabad, pp. 744–751, 2022.
27. Z. Gao, D. Zhang, J. Zhang, L. Liu, D. Niyato, and V. C. M. Leung, "World state attack to blockchain based IoV and efficient protection with hybrid RSUs architecture," *IEEE Transactions on Intelligent Transportation Systems*, vol. 24, no. 9, pp. 9952–9965, Sept. 2023.
28. Z. Li, Y. Kong, C. Wang, and C. Jiang, "DDoS mitigation based on space-time flow regularities in IoV: a feature adaption reinforcement learning approach," *IEEE Transactions on Intelligent Transportation Systems*, vol. 23, no. 3, pp. 2262–2278, March 2022.
29. T. Zhang, C. Xu, P. Zou, H. Tian, X. Kuang, S. Yang, L. Zhong, and D. Niyato, "How to mitigate DDoS intelligently in SD-IoV: a moving target defense approach," *IEEE Transactions on Industrial Informatics*, vol. 19, no. 1, pp. 1097–1106, Jan. 2023.
30. M. N. Aman, U. Javaid, and B. Sikdar, "A privacy-preserving and scalable authentication protocol for the internet of vehicles," *IEEE Internet of Things Journal*, vol. 8, no. 2, pp. 1123–1139, Jan. 15, 2021.
31. S. Anbalagan, G. Raja, S. Gurumoorthy, R. D. Suresh, and K. Dev, "IIDS: intelligent intrusion detection system for sustainable development in autonomous vehicles," *IEEE Transactions on Intelligent Transportation Systems*, vol. 24, no. 12, pp. 15866–15875, Dec. 2023.
32. H. Cheng, J. Yang, M. Shojafar, J. Cao, N. Jiang, and Y. Liu, "VFAS: reliable and privacy-preserving V2F authentication scheme for road condition monitoring system in IoV," *IEEE Transactions on Vehicular Technology*, vol. 72, no. 6, pp. 7958–7972, June 2023.
33. W. Li, C. Xia, C. Wang, and T. Wang, "Secure and temporary access delegation with equality test for cloud-assisted IoV," *IEEE Transactions on Intelligent Transportation Systems*, vol. 23, no. 11, pp. 20187–20201, Nov. 2022.
34. Y. Q. Huang, and R. Q. Hu, "A privacy-preserving scheme for location-based services in the internet of vehicles," *Journal of Communications and Information Networks*, vol. 6, no. 4, pp. 385–395, Dec. 2021.
35. N. U. Saqib, S. U. Malik, A. Anjum, M. H. Syed, S. A. Moqurrab, G. Srivastava, and J. C. W. Lin, "Preserving privacy in internet of vehicles (IoV): a novel group-leader-based shadowing scheme using blockchain," *IEEE Internet of Things Journal*, vol. 10, no. 24, pp. 21421–21430, Dec. 15, 2023.
36. C. Feng, K. Yu, M. Aloqaily, M. Alazab, Z. Lv, and S. Mumtaz, "Attribute-based encryption with parallel outsourced decryption for edge intelligent IoV," *IEEE Transactions on Vehicular Technology*, vol. 69, no. 11, pp. 13784–13795, Nov. 2020.
37. H. Karim, and D. B. Rawat, "TollsOnly please – homomorphic encryption for toll transponder privacy in internet of vehicles," *IEEE Internet of Things Journal*, vol. 9, no. 4, pp. 2627–2636, Feb. 15, 2022.
38. K. -Y. Lam, S. Mitra, F. Gondesen, and X. Yi, "ANT-centric IoT security reference architecture – security-by-design for satellite-enabled smart cities," *IEEE Internet of Things Journal*, vol. 9, no. 8, pp. 5895–5908, April 15, 2022.
39. C. Aristidou, and E. Marcou, "Blockchain standards and government applications," *Journal of ICT Standardization*, vol. 7, no. 3, pp. 287–312, 2019, doi: 10.13052/jicts2245-800X.736.
40. L. Yang, A. Moubayed, and A. Shami, "MTH-IDS: a multitiered hybrid intrusion detection system for internet of vehicles," *IEEE Internet of Things Journal*, vol. 9, no. 1, pp. 616–632, Jan. 1, 2022.
41. B. Lampe, and W. Meng, "Intrusion detection in the automotive domain: a comprehensive review," *IEEE Communications Surveys & Tutorials*, vol. 25, no. 4, pp. 2356–2426, Fourthquarter 2023.

42. R. Jabbar, N. Fetais, M. Kharbeche, M. Krichen, K. Barkaoui, and M. Shinoy, "Blockchain for the internet of vehicles: how to use blockchain to secure vehicle-to-everything (V2X) communication and payment?," *IEEE Sensors Journal*, vol. 21, no. 14, pp. 15807–15823, July 15, 2021.
43. L. Malina, P. Dzurenda, S. Ricci, J. Hajny, G. Srivastava, R. Matulevičius, A. A. O. Affia, M. Laurent, N. H. Sultan, and Q. Tang, "Post-quantum era privacy protection for intelligent infrastructures," *IEEE Access*, vol. 9, pp. 36038–36077, 2021.
44. Z. Wang, Z. Li, Z. Li, Y. Xu, X. Yang, F. Qi, and H. Jia, "FRNet: an MCS framework for efficient and secure data sensing and privacy protection in IoVs," *IEEE Internet of Things Journal*, vol. 10, no. 18, pp. 16343–16357, Sept. 15, 2023.
45. B. Riedl, V. Grascher, S. Fenz, and T. Neubauer, "Pseudonymization for improving the Privacy in E-Health applications," in *Proceedings of the 41st Annual Hawaii International Conference on System Sciences (HICSS 2008)*, Waikoloa, HI, USA, 2008, pp. 255–255.
46. A. Pawar, S. Ahirrao, and P. P. Churi, "Anonymization techniques for protecting privacy: a survey," in *2018 IEEE Punecon*, IEEE, Pune, India, 2018, pp. 1–6.
47. B. Liu, Z. Luo, H. Chen, and C. Li, "A survey of state-of-the-art on edge computing: theoretical models, technologies, directions, and development Paths," *IEEE Access*, vol. 10, pp. 54038–54063, 2022
48. H. Ji, O. Alfarraj, and A. Tolba, "Artificial intelligence-empowered edge of vehicles: architecture, enabling technologies, and applications," *IEEE Access*, vol. 8, pp. 61020–61034, 2020.
49. D. Das, S. Banerjee, P. Chatterjee, U. Ghosh, W. Mansoor, and U. Biswas, "Design of an automated blockchain-enabled vehicle data management system," in *2022 5th International Conference on Signal Processing and Information Security (ICSPIS)*, Dubai, United Arab Emirates, 2022, pp. 22–25.
50. V. C. Gupta, S. Gabadia, M. Agarwal, and K. Samdani, "An intrinsic review on securitization using blockchain," *2021 International Conference on Computational Performance Evaluation (ComPE)*, Shillong, India, 2021, pp. 971–976.
51. S. Ma, S. Wang, and W. -T. Tsai, "Delay analysis of consensus communication for blockchain-based applications using network calculus," *IEEE Wireless Communications Letters,* vol. 11, no. 9, pp. 1825–1829, Sept. 2022.
52. S. Yu, J. Lee, K. Park, A. K. Das, and Y. Park, "IoV-SMAP: secure and efficient message authentication protocol for IoV in Smart City environment," *IEEE Access*, vol. 8, pp. 167875–167886, 2020.
53. A. Majeed, and S. O. Hwang, "A comprehensive analysis of privacy protection techniques developed for COVID-19 pandemic," *IEEE Access*, vol. 9, pp. 164159–164187, 2021.
54. R. Kumar, P. Kumar, R. Tripathi, G. P. Gupta, and N. Kumar, "P2SF-IoV: a privacy-preservation-based secured framework for internet of vehicles," *IEEE Transactions on Intelligent Transportation Systems*, vol. 23, no. 11, pp. 22571–22582, Nov. 2022.
55. R. Silva, and R. Iqbal, "Ethical implications of social internet of vehicles systems," *IEEE Internet of Things Journal*, vol. 6, no. 1, pp. 517–531, Feb. 2019.
56. M. Ramalingam, G. C. Selvi, N. Victor, R. Chengoden, S. Bhattacharya, P. K. R. Maddikunta, D. Lee, M. J. Piran, N. Khare, G. Yenduri, and T. R. Gadekallu, "A comprehensive analysis of blockchain applications for securing computer vision systems," *IEEE Access*, vol. 11, pp. 107309–107330, 2023, doi: 10.1109/ACCESS.2023.3319089.
57. S. M. Karim, A. Habbal, S. A. Chaudhry, and A. Irshad, "BSDCE-IoV: blockchain-based secure data collection and exchange scheme for IoV in 5G environment," *IEEE Access*, vol. 11, pp. 36158–36175, 2023.
58. S. Ma, S. Wang, and W. -T. Tsai, "Delay analysis of consensus communication for blockchain-based applications using network calculus," *IEEE Wireless Communications Letters*, vol. 11, no. 9, pp. 1825–1829, Sept. 2022.

59. Z. Bao, Q. Wang, W. Shi, L. Wang, H. Lei, and B. Chen, "When blockchain meets SGX: an overview, challenges, and open issues," *IEEE Access*, vol. 8, pp. 170404–170420, 2020.
60. S. Islam, M. J. Islam, M. Hossain, S. Noor, K. S. Kwak, and S. M. R. Islam, "A survey on consensus algorithms in blockchain-based applications: architecture, taxonomy, and operational issues," *IEEE Access*, vol. 11, pp. 39066–39082, 2023.

2 Smart Mobility
Cognitive Computing in Modern Transportation Systems

S. Delsi Robinsha, B. Amutha, D. Vanusha, and D. Vathana

2.1 INTRODUCTION

The transport industry is currently undergoing a transformational evolution as a result of the incorporation of cutting-edge technologies that are designed to handle the myriad of difficulties that are associated with urban mobility. Cognitive computing, a subfield of artificial intelligence (AI) that focuses on the development of self-learning systems through the use of data mining, pattern recognition, and natural language processing, is at the vanguard of this revolution. The application of cognitive computing is essential for the development of smart mobility solutions that improve the effectiveness, safety, and sustainability of transportation systems. This is because cognitive computing enables intelligent decision-making and adaptive responses.

2.1.1 Smart Mobility: A General Overview

Smart mobility is the practice of enhancing transport systems by the integration of new and improved technology in order to make them more efficient, environmentally friendly, and user-friendly. Personalised travel services, real-time traffic management, driverless cars, and intelligent transportation systems (ITS) are all part of it. Congestion reduction, environmental impact mitigation, safety enhancement, and user experience improvement are the main objectives of smart transportation. The concept of "smart mobility" refers to the use of cutting-edge technologies such as artificial intelligence (AI), machine learning (ML), and the Internet of Things (IoT) to build transportation networks that are reactive to current situations and consumer demands.

2.1.2 The Significance of Cognitive Computing Systems in the Transportation Sector

In order to analyse massive amounts of data, recognise patterns, and make decisions based on that analysis, cognitive computing is an essential component in the development of smart mobility. It does this by providing the required tools and frameworks.

DOI: 10.1201/9781003392330-2

The following are some of the most important aspects of its significance in the transportation sector:

- Cognitive computing systems are able to process and analyse real-time data from a variety of sources, including sensors, cameras, and GPS devices, in order to generate insights that may be put into action for the purpose of traffic management, route optimisation, and emergency response.
- Cognitive computing has the ability to foresee probable breakdowns and maintenance needs by continuously monitoring the health of transportation infrastructure and vehicles. This allows for a reduction in both downtime and operational expenses, also known as predictive maintenance.
- Cognitive computing technologies are essential for the development of autonomous vehicles because they allow for the interpretation of sensory input, the making of driving judgements, and the efficient and safe navigation of complicated surroundings.
- Cognitive computing makes it possible to personalise travel alternatives by taking into account individual preferences, historical data, and current conditions. This not only improves the user experience but also encourages the use of public transit (also known as "public transportation").
- Cognitive computing has the ability to identify potential safety concerns and security threats through the use of advanced analytics and pattern recognition. This enables proactive steps to be taken to improve safety and security.

2.1.3 The Chapter's Purpose and Scope

The primary topics covered in this chapter are smart mobility and the revolutionary impact of cognitive computing on transportation systems. As a result, research on the underlying technology, key applications, pros, cons, and growth potential is required, in order to lay out all the groundwork for understanding smart mobility and how cognitive computing has contributed to its growth. The goal of this chapter is to:

- Learn more about AI, ML, the IoT, and real-time data analytics as they pertain to transportation [1].
- Examine how cognitive computing has affected important domains like AVs, traffic control, predictive maintenance, and tailored vacations.
- Talk about the things to think about and address when putting cognitive computing into transportation, such as data privacy, cybersecurity, and ethical concerns.
- Suggest a course of action for further study and development in the topic, drawing attention to current tendencies and future possibilities.

2.2 COGNITIVE COMPUTING TECHNOLOGIES

ML looks at data and finds patterns and conclusions by using statistical models and algorithms. Without any help, it learns from new information and changes based on that. It gives computers the chance to learn from algorithms that have already been

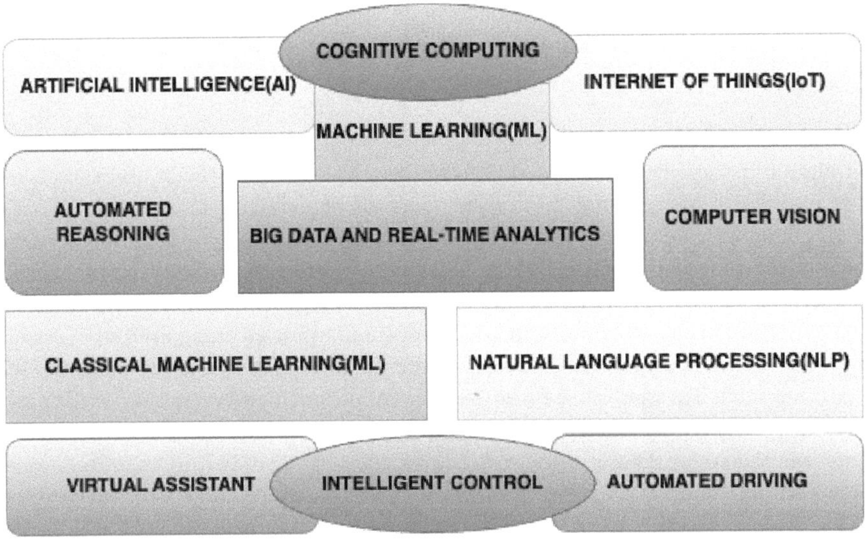

FIGURE 2.1 Cognitive computing technologies.

written. Also, AI gives computers the rules they need to work smartly. Machines can solve hard problems and make choices that give them the best chance of success because they are smart. Cognitive computing is an autonomous system that learns on its own and uses algorithms for ML and data mining, neural networks, and visual recognition to do smart jobs that humans would normally do [2]. The goal of cognitive computing is to solve hard problems by imitating how people think and act. It learns on a large scale, thinks carefully, and easily interacts with people. Neural networks and deep learning are the main parts of cognitive computing. Visual recognition looks at trends in a picture or video and figures out what it is by using deep learning and neural network algorithms. It looks through pictures to find scenes, items, text, and other things. The Google Lens app is a good example of visual recognition because it uses the camera on our phones to take pictures of things and tell us about them. The skill of being able to understand human words is called natural language processing (NLP). It also knows and processes a lot of natural language data and looks at them to draw conclusions. The smart compose feature of Gmail, which tells you what words and sentences to write next, is a very popular example of NLP. When you're done writing an email, it tells you what kind of mood it was written in, like formal, offensive, appreciative, or anger (Figure 2.1).

2.2.1 Artificial Intelligence

A collection of technologies collectively known as artificial intelligence allows computers to do a wide range of complex tasks, such as visual perception, language comprehension and translation, data analysis, suggestion making, and much more. When it comes to cutting-edge computing, AI is the key that opens all the doors for consumers and companies alike. By applying AI to photographs and documents, optical character recognition (OCR) can transform unstructured content into structured data that are ready for business use, allowing for the extraction of text and data as well as the unlocking of important insights [3].

2.2.1.1 Autonomous Driving

AI algorithms integrate sensory data from cameras, LiDAR, radar, and other sensors in order to navigate and control autonomous vehicles. This method is referred to as autonomous driving. Autonomous cars are beginning to emerge as a viable option in certain sectors of the industrial sector. There are many examples, including agriculture, transportation, and the military. As time goes on, we are getting closer and closer to the day when we will see autonomous vehicles being used in everyday life by the average consumer. The information that is gathered from sensors and the algorithms that are used by AI are the foundation for many of the tasks that cars must carry out. It is necessary for vehicles to gather data, plan their course, and then carry out the route. These tasks, particularly the final two, call for non-traditional programming approaches and rely on ML techniques, which are a component of AI.

2.2.1.2 Traffic Management

AI is a specific field within computer science that concentrates on developing computers with the ability to carry out tasks that usually necessitate human intelligence. AI utilises a range of methodologies, such as ML and deep learning, to facilitate the acquisition of knowledge from data and facilitate decision-making or prediction-making processes [4]. AI has the capability to rapidly analyse current data and make immediate decisions, resulting in smooth traffic flow and prompt reactions to issues. AI improves traffic signal timings, provides optimal route ideas, and manages lanes to minimise congestion and decrease travel durations. AI systems have the capability to identify accidents, instances of reckless driving, and other risks, hence improving road safety for both motorists and pedestrians. AI aids in the reduction of fuel usage and greenhouse gas emissions by minimising stop-and-go traffic and optimising routes. AI implemented in traffic management systems has the potential to enhance urban mobility, hence increasing the accessibility of transportation. Entrepreneurs seeking to enter this dynamic field may find great potential in investigating different small company concepts related to AI and traffic management.

2.2.1.3 Predictive Analytics

Supply chain workers in the transportation business have access to a wide variety of technologies that are meant to assist them in processing the massive amounts of data that they receive on a daily basis. By incorporating predictive analytics into your transportation management process, you can improve your ability to anticipate service failures before they occur. It is possible for service failures to trigger a line or even a plant shutdown, which would result in extremely high costs, if a product is going to be manufactured in a facility that uses a just-in-time (JIT) inventory environment. It is possible for customers to suffer significant fees as a result of service failures and late loads while they are on their way to a big-box distribution centre. These fines could build up over time and have a negative impact on the bottom line.

2.2.2 Machine Learning

The field of study known as machine learning is responsible for providing computers with the ability to learn without being explicitly programmed. ML is without a doubt one of the most fascinating technologies that one may come upon [5]. As is obvious from the name, it provides the computer with characteristics that make it more comparable to human beings.

2.2.2.1 Anomaly Detection

ML anomaly detection is incredibly useful in transportation applications because it can detect suspicious security threats, broken equipment in cold chain logistics, unusual traffic patterns, and unusual vehicle movements in real time. This allows transportation authorities and operators to respond swiftly, which improves safety, security, and operational efficiency across all modes of transportation. Automated analysis of video footage, sensor data, and other transportation-related information by ML models allows for the reliable detection of out-of-the-ordinary occurrences, changing the game for transportation systems in terms of monitoring, management, and optimisation of operations.

2.2.2.2 Route Optimisation

The transportation and logistics industries are witnessing a paradigm shift in route optimisation as a result of ML. This new approach leverages real-time data analysis, predictive analytics, capacity planning, dynamic routing, and predictive maintenance to drastically improve efficiency and cut costs. With the use of real-time data on vehicle locations, fuel consumption, and delivery times, ML algorithms can optimise routes in real time and allocate orders to maximise fleet utilisation. They can also analyse past data to estimate future demand and traffic patterns. Logistics routes can be better planned with the help of ML models by anticipating maintenance needs and vehicle capacity, which in turn improves delivery times, decreases transportation costs, and makes customers happier [6,7]. The logistics and transportation industries stand to gain even more from the optimisation and automation made possible by ML's superior analytical capabilities as these systems keep learning from more data over time.

2.2.2.3 Demand Forecasting

ML is changing the game when it comes to transportation and logistics demand forecasting. With the help of historical data, market trends, and external elements like weather and seasons, firms can now reliably predict future demand patterns. Optimising inventory levels, reducing stockouts, and improving customer happiness by ensuring products are accessible when needed are all possible thanks to ML algorithms' ability to analyse massive volumes of data, identify hidden relationships, and create precise forecasts. In addition to improving operational efficiency, ML-powered demand forecasting allows dynamic route optimisation by incorporating real-time data on traffic, delivery windows, and package characteristics, which in turn reduces fuel costs and delivery times. The logistics business is being revolutionised by ML, which can automate forecasts and manage complicated statistics. This is making the industry more intelligent, responsive, and sustainable (Figure 2.2).

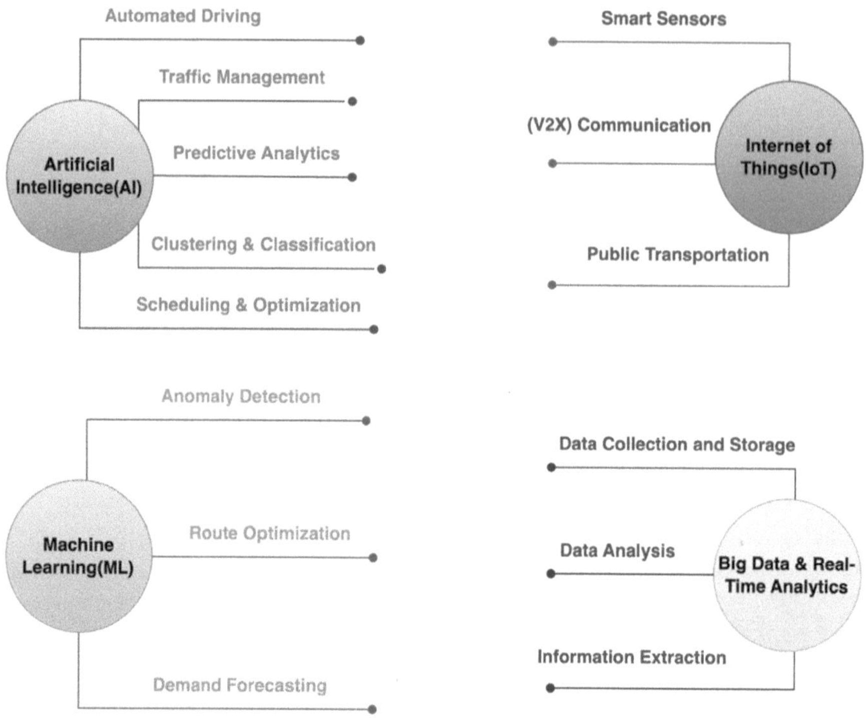

FIGURE 2.2 Interconnections and applications of cognitive computing.

2.2.3 Internet of Things

Much acclaim goes to the ground-breaking use of the IoT in transit for making our lives easier and more streamlined. For both our daily commute and international logistics, this game-changing technology ushers in a new era of connectedness, safety, and efficiency. Transportation applications that leverage the IoT are helping to smarten cities and make city processes more controllable, thanks to technological developments in both fixed and mobile apps. With cellular networks now being as reliable as old wired networks in fixed applications like cameras, traffic lights, and junction management, there are almost no opportunities left and the deployment is outrageously expensive [8]. With the introduction of 5G and the fast expansion of 4G networks, mobile router technology has reached a point where networks are efficiently fast for applications like paratransit vehicles, buses, light rail, ambulances, and police and fire departments.

2.2.3.1 Smart Sensors

The IoT devices that offer real-time data on the location, speed, and conditions of a vehicle include GPS trackers, accelerometers, and environmental sensors. Transportation is changing because smart sensors give real-time info for safety and efficiency. They find obstacles, keep track of speeds, and check on the state of infrastructure. This makes it possible for things like self-driving cars, traffic lights that change based on traffic, and planned repair. Their operations and customer service are better because they can keep track of goods and fleets. As tracking technologies get better, they will keep making transportation systems safer, faster, and smarter.

2.2.3.2 Vehicle-to-Everything (V2X) Communication

The IoT makes it possible for vehicles to interact with one another (V2V), with infrastructure (V2I), with pedestrians (V2P), and with networks (V2N), which improves transportation coordination and safety. V2X communication lets cars share real-time information with their surroundings, which makes things safer, more efficient, and better for the environment. Using wireless standards, V2X shares information about the state and dangers of vehicles. This makes it possible for applications like autonomous driving and avoiding collisions to work. It will be necessary for self-driving cars in the future, and it will also help traffic move and be better for the environment [9].

2.2.3.3 Public Transportation

By delivering real-time updates on schedules, delays, and congestion levels, the IoT will improve the efficiency and dependability of public transit.

2.2.4 Big Data and Real-Time Analytics

The transportation industry is undergoing a sea change due to the revolutionary insights into travel habits, traffic patterns, and operational efficiency made possible by big data and real-time analytics. With transportation data analytics, experts may access up-to-date information for any route in the world in a flash since it compiles data from several sources to provide detailed trip information. With big data as a foundation, transport experts can assess project outcomes and refine their forecasts. Big data facilitates easier air travel by assisting in the effective management of flights, passengers, and airport resources, while smart cities utilise ML to assess traffic data and develop real-time traffic management systems. Freight and logistics plans and routes can also be enhanced with data analytics, leading to faster delivery times and more efficiency overall. The transportation industry is about to undergo a radical transformation due to the exponential growth of these technologies [10].

2.2.4.1 Data Collection and Storage

Real-time analytics and big data are changing transportation by giving us new ways to look at travel trends, traffic flows, and how well things work. Transportation data analytics are important ways to gather and store data because they combine data from GPS, cell phone signals, and sensors to give detailed information on trips, routes, speeds, durations, modes, and more. Automatic toll collection systems can also be used to find vehicles and send real-time information about traffic jams and delays. Permanent traffic tracking systems use inductive loops, piezo sensors, and video imaging to collect long-term traffic data. Video imaging technology that uses "smart" cameras to find and sort cars also gives good traffic information. The information gathered in this way is then saved and analysed so that it can be used for things like managing traffic, finding the best routes, organising airports, and planning and modelling transportation [11,12].

As these technologies improve, transportation agencies will be able to make better choices based on data, which will make transportation systems more efficient and effective.

2.2.4.2 Data Analysis

Big data and real-time analytics in transportation depend on data processing. To give professionals quick access to correct data, transportation data analytics combine data from different sources to give detailed information on trips, routes, speeds, and modes. Environmental isotopes can be used in advanced methods, such as atom trap trace analysis, to find out how the oceans, mountain ice, and groundwater move. Real-time analytics combined with the IoT make applications like driverless cars and traffic control possible. AI-powered video analysis makes railway operations safer and smarter. Transportation agencies can make better choices about how to run their businesses and maintain their infrastructure as these data analysis tools get better.

2.3 INTELLIGENT TRANSPORTATION INFRASTRUCTURES

The term "intelligent transportation systems" (ITS) refers to the integration of various management techniques, as well as computer, electronic, and communication technologies, in order to deliver information to travellers to improve the highway transportation systems' safety and efficiency. User services, ITS architecture, and ITS planning are the primary topics that are covered in this chapter. Numerous user services provided by ITS are categorised into eight categories, and each of these categories has been briefly defined. A brief description of the ITS design, which emphasises both logical and physical architecture, is provided here. This architecture offers a standard framework for planning, specifying, and integrating intelligent transportation systems.

By facilitating the integration of sophisticated communications technologies into the transportation infrastructure and into vehicles, ITS increase transportation safety and mobility, as well as boosts global connectivity. This is accomplished through the enhancement of levels of productivity. For the purpose of improving traffic management and making the most of the transportation infrastructure that is already in place, ITS incorporate a wide variety of information and electronic technologies that are based on wireless and wired communication to maximise utilisation. Because of this, the driving experience, the capacity of road networks, the reduction of dangers in transportation, the alleviation of traffic congestion, the improvement of transportation efficiency, and the reduction of pollution are all enhanced.

2.3.1 Overview of Intelligent Transportation Systems

The goal of implementing ITS is to make transportation networks more efficient, safer, and environmentally friendly by integrating various forms of information and communication technology into transportation infrastructure and vehicles. ITS include many different kinds of technology and uses them for many different things, such as (Figure 2.3):

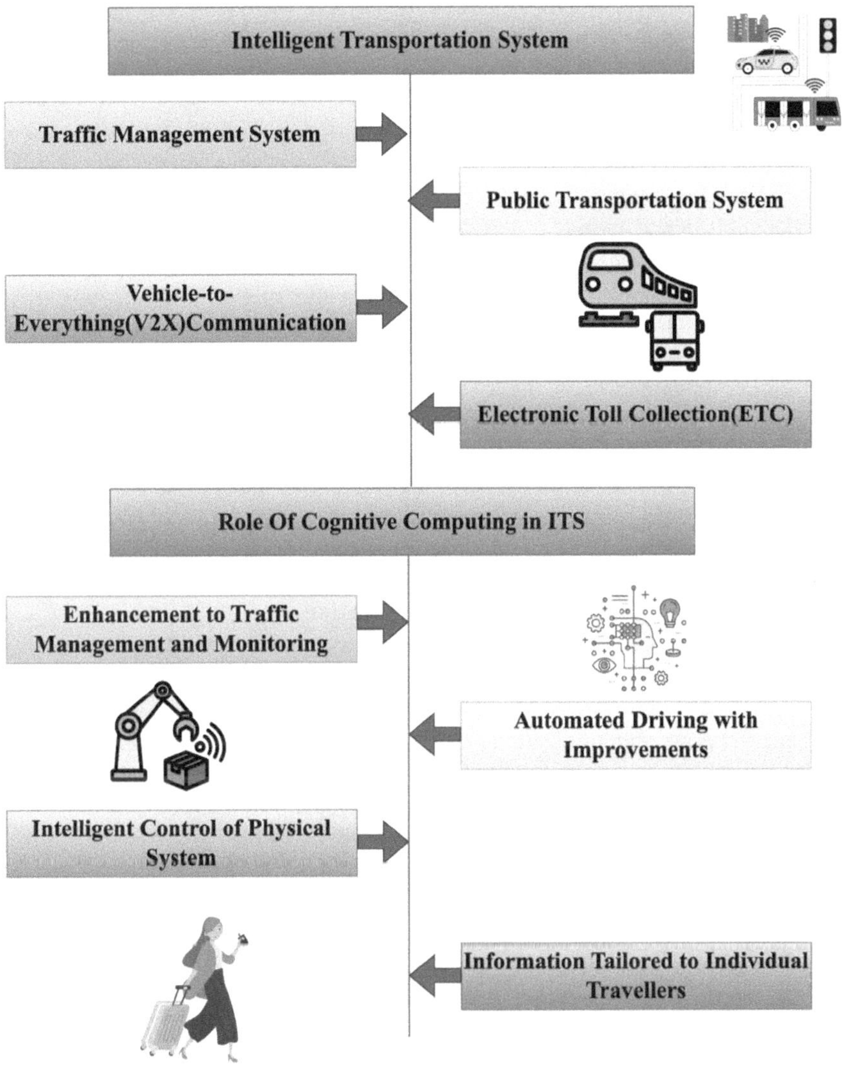

FIGURE 2.3 Intelligent transportation infrastructures.

A. **Traffic Management Systems**:

Traffic management systems (TMS) actively monitor, optimise, and respond to traffic conditions in real time by using sensors, signals, and control centres. Detectors that record traffic flow, variable message signs that educate drivers, and ramp metres that control vehicle entry are essential parts. TMS features centre on coordinating emergencies, optimising flows using tactics like ramp metering, and detecting incidents. Reduced congestion, increased safety, effective infrastructure utilisation, and lower environmental impact are just a few of the considerable benefits that may be achieved by using these capabilities of TMS. Greater optimisation will be possible through integration with connected and autonomous vehicle systems as transportation technology advances and TMS capabilities expand.

B. **Public Transportation Systems**:

Buses, trains, subways, and light rail are all part of public transport systems, which are shared passenger transport services that the general public can use. Reduced traffic congestion and environmental effect, enhanced accessibility and mobility, individual cost savings, economic development support, quality of life improvements, and social fairness are just a few of the many community benefits that these systems offer. A public transport system primarily consists of vehicles, routes and schedules that are set in stone, stations and terminals, methods for collecting fares, and facilities for operations and maintenance. Commuter patterns and travel habits are changing, so is the way public transportation is adapting to accommodate them. Other important trends include the use of technology for real-time tracking and mobile ticketing, the growth of high-capacity modes like light rail and bus rapid transit, the integration of different forms of transportation, the emphasis on sustainability through electrification, and the integration of other modes of transportation [13].

C. **Vehicle-to-Everything (V2X) Communication**:

V2X enables vehicles to exchange real-time data with their environment to improve safety, efficiency, and sustainability. Using wireless standards like DSRC and C-V2X, V2X shares information on vehicle status and hazards, powering applications such as collision avoidance, adaptive traffic signals, and autonomous driving. Key benefits include enhanced safety, optimised traffic flow, reduced environmental impact, and convenience for drivers. As autonomous vehicles advance, V2X will be critical for enabling safe self-driving capabilities. Challenges around security and standardisation need to be addressed for widespread adoption. Overall, V2X is revolutionising transportation by allowing vehicles to perceive and coordinate with their surroundings.

D. **Electronic Toll Collection**:

Electronic toll collection (ETC) is a computerised method that lets cars pay tolls without having to stop at a toll booth. Radio frequency identification (RFID) technology is used in ETC systems. Each car has a small RFID transponder or tag, which is usually attached to the windscreen. At toll booths, RFID readers can find and identify the transponders on cars that are coming up behind them. The reader sees the transponder and takes the right amount of money from the user's pre-paid account when the car goes through the toll plaza. As a backup, some ETC systems also use cameras that read licence plates to find cars that don't have a tracker. The main benefits of ETC are less traffic and travel times at toll plazas because vehicles can stay on the highway speed limit, better traffic flow and lower emissions from idling vehicles, more money for toll agencies because fewer people avoid paying the toll, less hassle for drivers because they don't have to stop and pay cash, and lower operating costs for toll agencies compared to manual toll collection. The northeastern United States has E-ZPass, California has FasTrak, Florida has SunPass, Germany has Toll Collect, and Portugal has Via Verde. These are some of the most well-known ETC systems. As ETC technology keeps getting better, it is becoming an important part of smart transportation systems that make traffic move better and lessen congestion.

2.3.2 Role of Cognitive Computing in ITS

Cognitive computing is a game-changer for ITS because it allows for more sophisticated data analysis, decision-making, and automation. Brain computing is revolutionising ITS in the following important ways:

A. **Enhancements to Traffic Management and Monitoring:**
   Traffic sensors, cameras, and linked vehicles are just a few examples of the many real-time data sources that cognitive systems may sift through. As a result, traffic incidents can be better detected, and proactive traffic management can be implemented. The ability to recognise trends, anticipate traffic jams, and optimise traffic lights and routing in real time is a remarkable capability of cognitive systems [14].
B. **Automated Driving with Improvements**:
   Autonomous vehicle technology relies heavily on cognitive computing. Cognitive systems are able to observe their environment, foresee any dangers, and make split-second judgements to navigate safely by combining input from sensors, mapping information, and ML algorithms. Because of this, self-driving cars are more dependable and have better decision-making abilities.
C. **Intelligent Control of Physical Systems**
   Roads, bridges, and signalling systems are all part of the transportation infrastructure that cognitive systems can keep an eye on. Problems can be located, maintenance needs can be anticipated, and infrastructure use can be optimised by analysing data collected by sensors. Because of this, data-driven, proactive infrastructure management is possible, which boosts efficiency and security.
D. **Information Tailored to Individual Travellers**
   Individual commuters can get tailored, up-to-the-minute travel advice and information from cognitive computing. Cognitive systems are able to provide personalised route guiding, multimodal trip planning, and predicted arrival times by analysing user preferences, traffic patterns, and contextual factors.

2.3.3 Case Studies of Intelligent Transportation Infrastructures

The revolutionary effect that ITS and cognitive computing have on transport systems is shown by its use in the real world. Here are a few noteworthy examples:

A. **Singapore's Electronic Road Pricing (ERP) System**
   Vehicles entering crowded areas during peak hours are charged a price through Singapore's ERP system, which is an ETC system. In order to deduct

the correct toll, it employs RFID readers mounted on overhead gantries to identify in-vehicle units. An integral part of Singapore's all-encompassing plan for managing transport demand is the ERP system. It has helped alleviate downtown traffic congestion by reducing the number of cars on the road during rush hour [15]. Toll rates are regularly changed by the dynamic ERP system in response to real-time traffic circumstances to optimise throughput. It has improved the use of public transit and helped decrease congestion by an estimated 20%–30%.

B. **London's Congestion Charging Scheme**

A road pricing system, the congestion charging plan in London levies a daily cost to vehicles that go into central London. The device enforces the charge using automated number plate recognition cameras. The plan was first implemented in 2003 with the intention of enhancing bus services, decreasing traffic, and raising funds for transportation infrastructure projects. Traffic entering the pricing zone has decreased by 15%–20%, bus speeds have improved, and important infrastructure projects have been supported by the cash earned, and hence, it has achieved its aims. The programme is adaptable, with daily charges changed in response to variables like inflation and traffic conditions [15].

C. **Stockholm's Congestion Tax**

In 2007, Stockholm imposed a congestion tax on vehicles that enter or leave the city centre. To collect the money, the government uses automated systems that detect licence plates. It has proved successful in increasing the efficiency of public transit and decreasing traffic by about 20%. All of the money that was collected from the congestion levy has been put back into the transport system. The system in Stockholm has time-based pricing, meaning that rates are greater during peak hours. Supporting Stockholm's objective of becoming fossil fuel-free by 2040, the congestion fee has been credited with decreasing emissions and improving air quality.

2.4 ADVANCED TRAFFIC MANAGEMENT SYSTEMS

In order to optimise urban traffic flow, decrease congestion, and improve road safety using real-time data, predictive analytics, and adaptive technology, advanced traffic management systems (ATMS) are crucial.

2.4.1 Real-Time Traffic Monitoring and Control

Using a variety of sensors and technology, real-time traffic monitoring and control entails constantly observing and managing traffic situations. Vehicle location, traffic conditions, speed, and volume are all tracked in real time by sensors and cameras. Decisions are made based on analyses of these data transmitted to centralised traffic control centres. Operators update drivers in real time on traffic, road closures, and alternate routes via dynamic message signs (DMS). In the event of an accident or a stranded vehicle, incident detection and management systems will immediately notify operators and emergency personnel. Furthermore, ramp metering regulates the pace at which vehicles access highways, smoothing traffic flow and avoiding congestion, and variable speed limits change in real time depending on traffic circumstances to enhance safety and flow.

2.4.2 Traffic Prediction Systems

To better manage traffic, predictive traffic modelling takes both past and present data and uses them to make predictions about the future. In order to comprehend traffic patterns, this method entails collecting data from several sources, such as traffic sensors, GPS devices, and historical records. Algorithms trained on historical data can use this information to better forecast traffic situations in the future. The effect on traffic flow of various traffic management techniques, including road closures or adjustments to traffic signals, can be evaluated through scenario simulations [16]. To help in planning and allocating resources, demand forecasting makes traffic predictions depending on things like time of day, weather, and special events. In addition, by identifying possible congestion points and times, predictive models enable pre-emptive actions to alleviate traffic jams and provide incident predictions, which in turn allow for quicker response and management.

2.4.3 Systems for Adaptive Traffic Signals

To optimise traffic flow and minimise delays, adaptive traffic signal systems (ATSS) dynamically change the timing of traffic signals based on real-time data. Detectors installed at junctions record the presence and movement of vehicles in real time, and traffic signal controllers use these data to control the operation of the signals. By analysing these data, adaptive algorithms can prioritise various traffic patterns and change the timing of signals accordingly. Vehicles are able to pass through numerous junctions without halting thanks to these systems, which synchronise traffic signals along corridors to generate green waves. Incorporating sensors that guarantee safe and efficient crossing times, ATSS also take pedestrians and cyclists into consideration. In times of crisis, these cars can take control of the traffic signals thanks to emergency vehicle pre-emption. The dependability and efficiency of public transport services are enhanced through the practice of public transport prioritisation, which grants buses and trams preference at traffic lights. Better urban mobility, less congestion, and safer roads are the results of intelligent, responsive transport networks made possible by integrating these technologies [17].

2.5 PREDICTIVE MAINTENANCE

Through the use of cutting-edge technology, predictive maintenance is able to foresee when equipment will break down, enabling repairs to be carried out prior to any downtime happening. Organisations may improve the dependability and efficiency of their transport systems by integrating sensors and the IoT, analysing data, and employing different predictive methodologies.

2.5.1 IoT and Sensor Integration

Integrating sensors and the IoT is essential to predictive maintenance because it provides the data needed to track and analyse equipment health in real time. Engines, brakes, tracks, and bridges are just a few examples of the essential infrastructure and vehicle parts that have smart sensors placed to continuously monitor factors like vibration, pressure, temperature, and wear. IoT devices gather massive volumes of data from these sensors and send them to centralised systems to be analysed. These data include operational conditions in real time as well as performance measurements from the past. With the advent of new technologies like 5G, the IoT has become an indispensable tool for the efficient transfer of data to remote servers or cloud storage. By doing preliminary data analysis locally and transmitting only essential information to central systems, edge computing reduces latency and speeds decision-making by processing data closer to its point of generation.

2.5.2 Analytics on Data for Predictive Mapping

Data analytics is essential for predictive maintenance because it takes raw sensor data and turns them into useful insights. Eliminating extraneous details, standardising formats, and fixing mistakes are all part of this data processing and cleaning process. Data trends and patterns can be better understood by applying statistical approaches; these methods also aid in detecting variations from typical operating circumstances that may indicate impending breakdowns [18]. In order to anticipate when machinery will break down, ML algorithms are taught to look for trends in both historical and real-time data. Various predictive techniques are employed to estimate the RUL of components and detect early indications of wear and tear, such as regression analysis, time-series analysis, and anomaly detection. In order to get a better picture of the equipment's health and any problems that may arise, visualisation tools like dashboards aid the maintenance staff in making sense of complicated data and trends.

2.5.3 Advantages and Efficient Use of Resources

Transportation networks may reap several benefits and save a tonne of money by switching to predictive maintenance. In order to keep vehicles and equipment running smoothly, organisations can reduce unplanned downtime by identifying and fixing problems before they fail. Equipment can last longer without experiencing major breakdowns thanks to preventative maintenance based on predictive insights. Because routine maintenance is usually cheaper than emergency repairs, predictive maintenance helps keep repair costs down. Transportation system safety is improved through early failure detection through the prevention of accidents caused by malfunctioning equipment. Scheduling maintenance more effectively allows for the reduction of needless routine checks and the reallocation of resources to areas that truly require them. Customer happiness and service quality are both enhanced by predictive maintenance's capacity to make transportation systems more reliable.

2.6 CHALLENGES AND CONSIDERATIONS

For effective acceptance and operation, cognitive computing implementation in transport systems brings a number of concerns and problems that must be fixed. Protecting personal information, navigating complex systems, meeting stringent operational requirements, and dealing with policy and regulatory concerns are all examples of the issues that may arise.

2.6.1 Data Security and Privacy Rule

Using cognitive computing systems in transportation must adhere to the highest standards of data privacy and security. Massive data sets, including personally identifiable information about people and their whereabouts, are produced by the widespread use of IoT devices and sensors. Strict access rules, safe data storage, and strong encryption technologies are necessary to protect these data's privacy [19]. Data anonymisation techniques can also be used to safeguard personal information. It is imperative to implement cybersecurity measures to safeguard transportation data against cyber threats such as hacking, data breaches, and other similar incidents. Threats and vulnerabilities are always changing, and thus, it's important to monitor and update security policies constantly.

2.6.2 Moral Issues with Cognitive Computing

Several moral questions arise from the potential use of cognitive computing in transportation systems. Decisions made by intelligent systems and autonomous cars can have far-reaching effects on human safety. Designing these technologies with human safety and ethical decision-making in mind is of the utmost importance. We must resolve concerns including the possibility of bias in AI systems, the lack of openness surrounding decision-making, and the lack of responsibility for mistakes or accidents. The development, implementation, and maintenance of transportation-related cognitive computing systems should be regulated by established ethical frameworks and standards that are in harmony with accepted social norms and principles.

2.6.3 Difficulties with Technology and Operations

Cognitive computing system implementation in transportation is fraught with operational and technical hurdles. Integrating many technologies like AI, ML, IoT, and big data analytics calls for a lot of knowledge and teamwork. For everything to work smoothly, it is crucial that various systems and equipment be able to communicate with one another. Cognitive computing systems can only function with trustworthy and precise data and sensors. Processing and analysing massive amounts of data in real time also require a lot of computing power, which calls for reliable infrastructure and effective algorithms [20]. To fix any technological difficulties that may develop and keep these systems running well, maintenance and upgrades are essential.

2.7 FUTURE DIRECTIONS AND INNOVATIONS

Cognitive computing has a huge amount of ability to make huge steps forward in transport, and a number of new trends are set to completely change the field. Better AI and ML algorithms are making prediction models smarter and more accurate, which helps people make better decisions in real time. When 5G technology is added, it speeds up data transfer, lowers latency, and improves connectivity. This makes it easier for IoT devices to work together and for real-time analytics to be done. Improvements in edge computing will make it possible for more data processing to happen closer to where the data come from, which will cut down on delay and bandwidth use. Blockchain technology is being looked into as a way to improve data security and accuracy by making records clear and impossible to change. The focus on working together between people and machines is also meant to boost the safety and performance of the whole system by using the best features of both human workers and cognitive systems. The transport industry is on the brink of a revolution due to a number of new developments in cognitive computing. Improved decision-making in real-time is a direct result of the rise of more complex and accurate predictive models made possible by advances in AI and ML. IoT devices and real-time analytics in transportation systems can function seamlessly with the help of 5G technology, which offers quicker data transmission, lower latency, and improved connectivity. Cognitive computing in transportation might be drastically altered by a number of new technologies. Complex transportation optimisation issues are currently unsolvable by classical computers, but quantum computing has the potential to change all of that. Cognitive computing systems' capacity for monitoring and prediction will be enhanced by the creation of more precise and sensitive sensors.

2.8 CONCLUSION

The integration of cognitive computing, AI, ML, IoT, and real-time analytics is crucial to the advancement of smart mobility, which aims to build transport systems that are more efficient, safe, and environmentally friendly. Some of the primary issues covered are how cognitive computing is going to revolutionise transportation systems, driverless cars, traffic management, predictive maintenance, and new developments in the future. Significant advantages, including less congestion, increased safety, a better user experience, and more operational efficiency, are offered by cognitive computing's influence on smart mobility. These innovations enhance transit networks while also lowering emissions and increasing resource efficiency, two overarching goals of sustainable urban development. To effectively harness the power of cognitive computing in transportation, we must tackle issues like data protection, ethical concerns, and legal frameworks. Urban areas must embrace these technologies if they are to become sustainable and robust enough to accommodate expanding populations and changing transportation demands.

REFERENCES

1. Sasikumar, A., Ravi, L., Devarajan, M., Kotb, H., & Subramaniyaswamy, V. (2024). Cognitive computing system-based dynamic decision control for smart city using reinforcement learning model. In R. Elakkiya & V. Subramaniyaswamy (Eds.), *Cognitive Analytics and Reinforcement Learning: Theories, Techniques and Applications* (pp. 29–50). John Wiley & Sons, Inc.
2. Ullah, U., Usama, M., Muhammad, Z., Akbar, A., Latif, S., & Ullah, R. (2024). Intelligent transportation channels for smart cities. In I. U. Khan, M. Ouaissa, M. Ouaissa, M. Fayaz, & R. Ullah (Eds.), *Artificial Intelligence for Intelligent Systems: Fundamentals, Challenges, and Applications* (pp. 280–323). CRC Press.
3. Dave Mahadevprasad, V., Rudhra, O., & Singh, S. K. (2024). Cognitive computing in smart cities and healthcare. In R. Elakkiya & V. Subramaniyaswamy (Eds.), *Cognitive Analytics and Reinforcement Learning: Theories, Techniques and Applications* (pp. 317–363). John Wiley & Sons, Inc.
4. Samson Arun Raj, A., & Yogesh, P. (2024). The future of modern transportation for smart cities using trackless tram networks. In Y. Wilks & T. Nishida (Eds.), *Conversational Artificial Intelligence* (pp. 369–384). CRC Press
5. Prakash, J., Murali, L., Manikandan, N., Nagaprasad, N., & Ramaswamy, K. (2024). A vehicular network based intelligent transport system for smart cities using machine learning algorithms. *Scientific Reports*, 14(1), p. 468.
6. Robinsha, S. D., & Amutha, B. (2024). Velocious: a resilient iot architecture for 6G based intelligent transportation system with expeditious movement mechanism. In *Wireless Personal Communications* (pp. 1–22).
7. Dureja, A., Dureja, A., Kumar, V., & Sabharwal, S. (2024). Combining digital twin technology and intelligent transportation systems for smart mobility. In V. Kumar & J. P. Liyanage (Eds.), *Transforming Industry using Digital Twin Technology* (pp. 281–296). Cham: Springer Nature Switzerland.
8. Xia, X., Lei, S., Chen, Y., Hua, S., & Gan, H. (2024). Highway smart transport in vehicle network based traffic management and behavioral analysis by machine learning models. *Computers and Electrical Engineering*, 114, p. 109092.
9. Jagatheesaperumal, S. K., Bibri, S. E., Huang, J., Rajapandian, J., & Parthiban, B. (2024). Artificial intelligence of things for smart cities: advanced solutions for enhancing transportation safety. *Computational Urban Science*, 4(1), p. 10.
10. Pradhan, D., Sahu, P. K., Tun, H. M., & Chatterjee, P. (Eds.). (2024). *Artificial and Cognitive Computing for Sustainable Healthcare Systems in Smart Cities*. John Wiley & Sons.
11. Li, Y., Lin, H., & Jin, J. (2024). Decision-making for sustainable urban transportation: a statistical exploration of innovative mobility solutions and reduced emissions. *Sustainable Cities and Society*, 102, p. 105219.
12. Shukla, R., Choudhary, A. K., Kumar, V. S., Tyagi, P., Mutharasan, A., Kumar, S., & Gupta, S. K. (2024). Understanding integration issues in intelligent transportation systems with IoT platforms, cloud computing, and connected vehicles. *Journal of Autonomous Intelligence*, 7(4), p. 1043.
13. Bijalwan, J. G., Singh, J., Ravi, V., Bijalwan, A., Alahmadi, T. J., Singh, P., & Diwakar, M. (2024). Navigating the future of secure and efficient intelligent transportation systems using AI and blockchain. *The Open Transportation Journal*, 18(1), p. e26671212291400.
14. Martí Gimeno, P. (2024). *Towards Sustainable and Efficient Road Transportation: Development of Artificial Intelligence Solutions for Urban and Interurban Mobility* (Doctoral dissertation), Universitat Politècnica de València, Valencia, Spain.
15. Hazarika, A., Choudhury, N., Nasralla, M. M., Khattak, S. B. A., & Rehman, I. U. (2024). Edge ML technique for smart traffic management in intelligent transportation systems. *IEEE Access* 12, pp. 25443–25458. https://doi.org/10.1109/ACCESS.2024.3365930.

16. Rehman, A., Haseeb, K., Alruwaili, F. F., Ara, A., & Saba, T. (2024). Autonomous and intelligent mobile multimedia cyber-physical system with secured heterogeneous IoT network. *Mobile Networks and Applications*, pp. 1–10.
17. Peldo, D., Banihashemi, S., LeNguyen, K., & Derrible, S. (2024). Navigating urban complexity: the transformative role of digital twins in smart city development. *Sustainable Cities and Society*, 111, p. 105583. https://doi.org/10.1016/j.scs.2024.105583
18. Valaskova, K., Nagy, M., & Grecu, G. (2024). Digital twin simulation modeling, artificial intelligence-based Internet of Manufacturing Things systems, and virtual machine and cognitive computing algorithms in the Industry 4.0-based Slovak labor market. *Oeconomia Copernicana*, 15(1), pp. 95–143.
19. Robinsha, S. D., & Amutha, B. (2023). IoT architecture for energy management in smart cities. *International Journal of Services Operations and Informatics*, 12(4), pp. 325–343.
20. Robinsha, S. D., & Amutha, B. (2023, November). IoT revolutionizing healthcare: a survey of smart healthcare system architectures. In *2023 International Conference on Research Methodologies in Knowledge Management, Artificial Intelligence and Telecommunication Engineering (RMKMATE)*, Chennai, India (pp. 1–5). IEEE.

3 Internal Combustion Engine Simulations in the Era of High-Performance Computing

Avinash Ravikumar and Benjamin Lawler

3.1 INTRODUCTION

Internal combustion engines (ICEs) have long been the dominant power source for automotive applications due to their high power and energy density. A major drawback of the ICEs is the use of fossil fuels, which can emit harmful greenhouse gases, detrimental to the environment, accelerating climate change. The research community continuously strives to optimize the ICEs for reduced emissions and increased efficiency. An in-depth understanding of the fundamental processes that occur in ICEs is essential for optimizing them.

Traditionally, the research community relies on experiments in physical prototypes to characterize ICEs and study the contributing factors. Nevertheless, experiments can be expensive, labor-intensive, and time-consuming. Therefore, researchers have developed mathematical models based on the acquired knowledge of the fundamental physics and chemistry to study ICEs. However, ICEs are complex since they involve multiple phenomena/processes and are highly unsteady in their operation. Thus, it requires multiple mathematical models to numerically simulate an ICE. These mathematical models are of varying fidelities, and the types of equations involved vary from simple algebraic equations to complex partial differential equations.

The accuracy of the solution and the computational expense/time are the two contradicting factors that are critical for determining which level of model fidelity is required for addressing a particular research question. The model choice depends on the tradeoff between required accuracy and affordable computational expense for the analysis. Moreover, understanding the limitations of a model and the nature of the process that the model tries to capture is necessary for extracting valuable information from a simulation of any fidelity. This chapter discusses the various fidelities of mathematical models or numerical tools employed to study ICEs. To provide a sense of the capabilities and limitations of a model, example case studies demonstrating the utilization of the models in ICE research are presented. This chapter discusses various mathematical models ranging from zero-dimensional algebraic equations to multi-dimensional partial differential equations, along with analyses performed with the corresponding models.

DOI: 10.1201/9781003392330-3

This chapter is structured hierarchically based on the computational expense of the models. The first section is dedicated to reduced-order models whose computational expense is relatively negligible with modern computers. The reduced-order models are further categorized as zero-dimensional and one-dimensional models, and an example case study is presented for each. Multi-physics simulations by synthesizing multiple reduced-order models are widely used in ICE research. A case study that combines the two reduced-order models is illustrated to conclude the first section and introduce the second section, which deals exclusively with multi-physics simulations.

The second section explains some of the important sub-models involved in a multi-physics simulation of an ICE with appropriate examples. Furthermore, two different system-level case studies are illustrated to highlight the potential of multi-physics simulations.

The third section will explore the high-fidelity three-dimensional models, which are computationally heavy and require high-performance computers for execution. Some physical sub-models are embedded in the three-dimensional simulations for efficient computation, described and discussed in the third section. The current state-of-the-art multi-physics simulations are performed by synthesizing high-fidelity models. A co-simulation methodology that combines two high-fidelity models is elaborated upon with an application case study to demonstrate this. The computational expenses of the various model fidelities are compared. Sometimes, the reduced-order models can be calibrated with information from high-fidelity simulations. A case study is presented to illustrate this calibration technique. Predominantly, the case studies presented involve heat transfer prediction since heat transfer in ICEs is one of the least understood processes due to its complexity. Finally, a separate section is dedicated to discussing the potential exploitation of the numerical tools to study advanced combustion modes, alternate fuels, and alternate ICE architectures. The structure of the chapter is as follows:

- Section-1: Reduced-order models
  - Zero-dimensional
  - One-dimensional
  - Reduced-order piston thermal analysis
- Section-2: Multi-physics simulation
  - Valve sub-model
  - Combustion sub-model
  - Thermal sub-model
  - System-level studies
- Section-3: High-fidelity simulation
  - Three-dimensional computational fluid dynamics
  - Finite element model
  - Co-simulation
- Computational expense
- Calibration of reduced-order model with high-fidelity simulation
- Numerical tools for assessing future ICEs

3.2 SECTION-1: REDUCED-ORDER MODELS

Most physical phenomena in an ICE have multiple degrees of freedom or contributing factors. It would be impossible to capture all the influencing factors in a single equation, and computational expense is a separate practical constraint. Therefore, assumptions are made based on available knowledge of a system or physical process to build mathematical models that can provide computationally efficient solutions for complex problems.

A system or physical process undergoes changes in time and space. The objective of a mathematical model is to capture changes (gradients) in both time and three-dimensional space. Suppose gradients in any of the dimensions are assumed to be negligible or unimportant from prior knowledge. In that case, the mathematical models can be reduced to solve for only the critical dimensions, commonly referred to as 'reduced-order' models. Nonetheless, the validity of the assumptions made in a reduced-order model should be carefully analyzed when applying to a problem to determine the solution's reliability. Despite this, reduced-order models are still widely used for their computational efficiency. Reduced-order models can be categorized based on the number of spatial dimensions considered (solved). Zero-dimensional and one-dimensional models are elaborated below, with an example case study for each.

3.2.1 ZERO-DIMENSIONAL (0-D) MODEL

A classic example of a 0-D model used by the ICE research community is the heat transfer correlations. Heat lost to the combustion chamber walls from the combustion gases is significant and irreversible. Thus, ample attention should be given to heat transfer to improve the efficiency of an ICE. Wall heat transfer in ICE is a complex phenomenon due to strong spatial gradients and the highly transient nature of internal thermodynamics and fluid motion. Direct measurements of heat flux to the combustion chamber walls are cumbersome due to mechanical limitations, particularly in moving components like the piston. Therefore, several researchers [1–5] in the past have attempted to develop 0-D mathematical models that could capture the surface-averaged transient heat loss through the combustion chamber walls. These mathematical models were developed using a multitude of techniques given as follows:

- Non-dimensional numbers such as Reynolds (Re), Prandtl (Pr), and Nusselt (Nu) numbers characterize flow inside the combustion chamber.
- Correlations developed from similar non-ICE experiments.
- Regression with experimental data from ICEs

A commonly used and simpler heat transfer correlation developed by Hohenberg shown in equation (3.1) is applied to a heavy-duty diesel engine operating at rated power to demonstrate the application:

$$h = 130\ V_c^{-0.06} p^{0.8} T^{-0.4} \left[\underline{v_p} + 1.4\right]^{0.8} \tag{3.1}$$

V_c – clearance volume, p – cylinder pressure, T – bulk gas temperature, $\underline{v_p}$ – mean piston speed

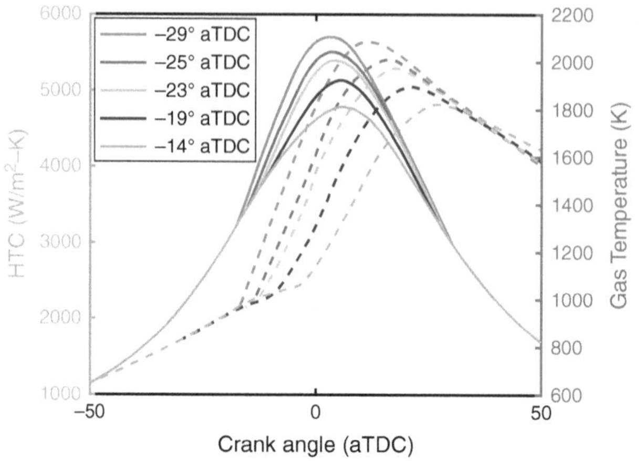

FIGURE 3.1 Heat transfer coefficients calculated using the heat transfer correlation and the gas temperatures for injection timing sweep at the rated power of a heavy-duty diesel engine.

The correlation requires cylinder pressure data collected from the experimental engine and the bulk gas temperature calculated from cylinder pressure using the ideal gas law. The gas temperatures and the calculated heat transfer coefficients for an injection sweep at the rated power condition of the heavy-duty diesel engine are plotted in Figure 3.1.

3.2.2 One-Dimensional (1-D) Model

One-dimensional (1-D) models capture a variable's gradient in the dimension of concern. 1-D models work well when the gradients in other dimensions are negligible compared to the relevant dimension. An example of such a model predominantly used in the ICE research is the 1-D unsteady heat conduction equation shown in equation (3.2):

$$\frac{d^2T}{dx^2} = \frac{1}{\alpha}\frac{dT}{dt} \tag{3.2}$$

T – temperature; α – thermal diffusivity; dx – distance; dt – change in time

An application of the equation is demonstrated here to predict the surface temperature of a combustion chamber component such as the piston. Commonly, the top side of the piston is exposed to the combustion gases, and the bottom side is cooled by oil. Thus, the heat conduction can be assumed to be dominantly in one direction from the piston's top surface to the bottom. Many studies have shown that the primary mode of heat transfer from combustion gases to the wall and the cooling side is convective [6]. Therefore, the boundary conditions for the top and the bottom of a piston are modeled using Newton's cooling law. On the other hand, Fourier's law of conduction says that the heat flux within a material is directly proportional to the gradient of the temperature in the normal direction. Equating Newton's cooling law and the Fourier law of conduction at the surface, equation (3.3) can be obtained:

$$q = h \cdot A \cdot \left(T_{\text{gas}} - T_{\text{wall}}\right) = -k \cdot \frac{\partial T}{\partial x}(x = 0) \quad (3.3)$$

q – heat flux; h – convective heat transfer coefficient; A – area; T_{gas} – gas temperature; T_{wall} – wall temperature

A pictorial representation of the 1-D piston model for piston surface temperature prediction using the 1-D unsteady heat conduction model is shown in Figure 3.2. The above-mentioned equation is continuous, and the equation should be discretized in time and space to be solved numerically. There are several methods available to discretize continuous equations for computationally efficient and stable solutions. Here, the 1-D unsteady heat conduction equation is discretized using second-order central spacing and forward difference time marching as shown in equation (3.4):

$$\frac{(T_{x+1,t-1} - 2.T_{x,t-1} + T_{x-1,t-1})}{\Delta x^2} = \frac{1}{\alpha}\frac{T_{x,t} - T_{x,t-1}}{\Delta t} \quad (3.4)$$

Similarly, the discretized equations for treating top and bottom surfaces considering the convective boundary conditions assuming unit area can be written as equation (3.5):

$$h \cdot \frac{T_{\text{gas},t} - T_{x=0,t}}{\Delta t} = -k \cdot \frac{T_{x,t} - T_{x+1,t}}{\Delta x} \quad (3.5)$$

An essential part of running transient simulations is choosing an optimal time step (Δt) for the selected grid size (Δx). From equation (3.4), a relation between the time step and the grid size can be derived ($\Delta t = \frac{\Delta x^2}{\alpha}$). This is the shortest time for the heat energy to flow from one node to another node for a material with its thermal diffusivity. Thus, the time step chosen must be shorter than this value to produce a stable and accurate solution. When a 5.5 mm thick piston crown made of martensitic steel is discretized by placing a node at every 0.01 mm, the maximum time step for a stable solution was calculated to be ~3.2 microseconds. When applying convective boundary conditions for both sides of the piston, the 1-D model could solve for piston temperature gradients in the normal direction that changes with time as plotted in Figure 3.3.

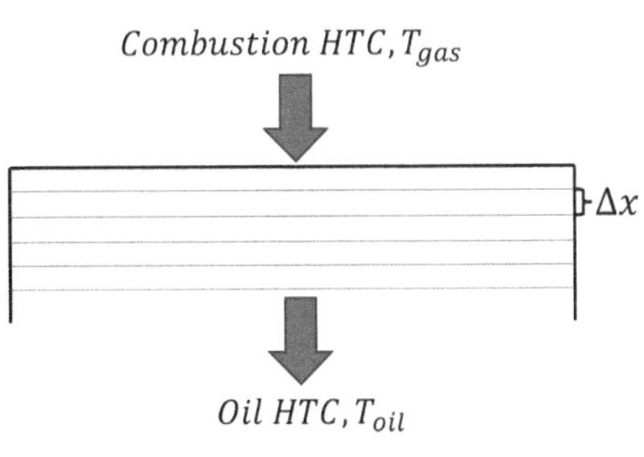

FIGURE 3.2 Pictorial representation of the piston discretized in 1-D and the boundary conditions, including the convective heat transfer coefficient (HTC) and fluid temperatures, for surface temperature prediction using 1-D heat conduction model.

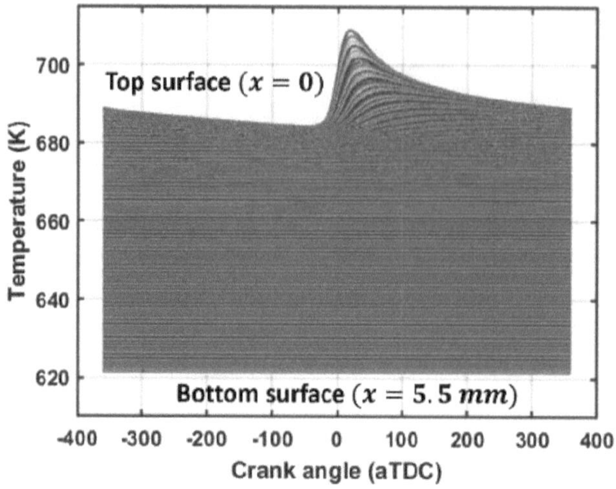

FIGURE 3.3 Transient nodal temperatures of a heavy-duty diesel engine piston predicted using the 1-D model.

0-D heat transfer correlations help estimate time-resolved heat transfer losses in an ICE. The 0-D heat transfer correlations can provide boundary conditions for estimating combustion chamber surface temperatures using the 1-D heat conduction model. This way of combining more than one mathematical model to capture multiple phenomena/processes simultaneously is known as 'multi-physics' simulations, which will be discussed in detail in the next section. A simple application of this concept is demonstrated below by performing heat transfer analysis of a heavy-duty diesel engine piston. Heavy-duty engines suffer from durability issues due to high thermal loads from combustion. Thermal analysis of the combustion chamber components could provide crucial insights to improve the engine's durability. Combining the 0-D heat transfer model and the 1-D heat conduction model discussed above could be an alternative to experiments that would be convenient for a preliminary thermal analysis of the piston.

3.2.3 Reduced-Order Piston Thermal Analysis

The heat transfer coefficients calculated using the 0-D Hohenberg heat transfer model and the gas temperatures in Figure 3.1 were applied as the combustion side boundary condition to the 1-D heat conduction model. The measured oil temperature and an assumed cooling heat transfer coefficient were used as the cooling side boundary condition to predict the surface temperature of the piston plotted in Figure 3.4 for an injection timing sweep at the rated power condition of the 1500 cc single-cylinder diesel engine [7]. The total combustion chamber heat loss as a percentage of input fuel energy and the maximum piston surface temperature for the injection timing sweep are plotted in Figure 3.5. It can be seen from Figure 3.5 that delaying the injection timing by 15 degrees reduced the total heat loss by approximately one percentage point and the maximum piston temperature by 63 K.

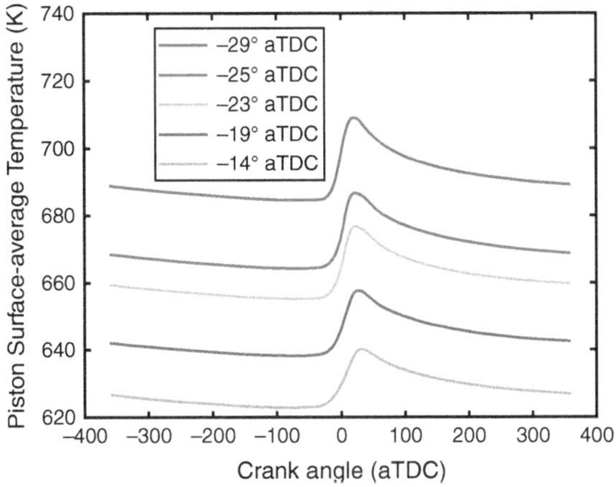

FIGURE 3.4 Piston surface-averaged temperatures predicted using the reduced-order thermal model for an injection timing sweep at rated power.

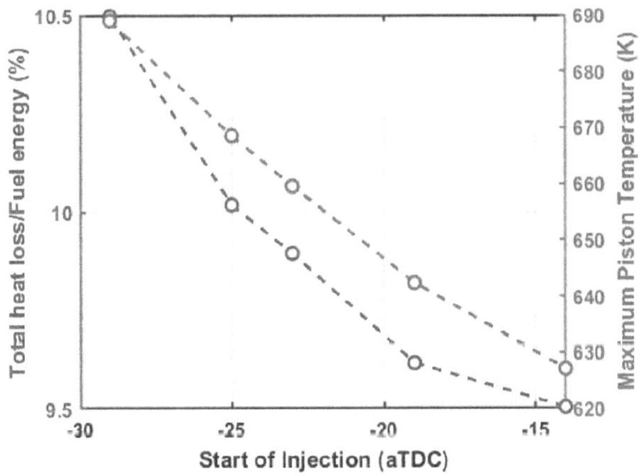

FIGURE 3.5 Total heat loss as a percentage of fuel energy and maximum piston temperature as a function of injection timing calculated using the reduced-order model.

The reduced-order piston thermal analysis performed by applying convective boundary conditions estimated from the 0-D heat transfer correlation on the 1-D heat conduction model helps predict heat loss and piston surface temperatures (an indication of thermal loading of the piston), as illustrated above. The above analysis could help engine calibration engineers to optimize the injection timing for best efficiency without exceeding the thermal limits of the piston. However, research [8] has shown that the 0-D heat transfer correlations showed significant discrepancies with measurements because of the assumptions. The foremost factor causing the discrepancy is the lack of spatial discretization since heat transfer in ICEs is spatially varying. Moreover, since the 0-D heat transfer correlation development relies on experimental data, they are restricted to those experimental conditions. On the other hand, 1-D heat conduction models are less reliable for complex geometries such as modern ICE pistons where lateral conduction is significant.

Multi-physics simulation tools such as GT-Suite can capture some level of spatial stratification with reduced-order models. Multi-physics simulations are inevitable for ICE research since multiple mediums are involved and multiple processes occur simultaneously. The reduced-order piston thermal analysis can be considered as a multi-physics simulation. Nonetheless, well-established multi-physics numerical simulation tools can simulate the entire engine from the intake air box to the tailpipe.

3.3 SECTION-2: MULTI-PHYSICS SIMULATION

The reduced-order simulations discussed above need measured data to perform the required analysis and lack accuracy due to the assumptions discussed previously. Multi-physics ICE simulation tools could perform a similar analysis without much use of experimental data by combining several models and removing some of the assumptions. ICEs are complex machines that involve the interaction of multiple mediums such as liquids, gases, and solids. Also, several physical or chemical phenomena such as combustion, and fluid flow can occur simultaneously and their effects on each other are coupled. It would be impossible to develop a single mathematical model to simulate an entire ICE.

GT-Suite is a simulation platform that couples multiple mathematical models for chemistry, flow, thermodynamics, mechanics, etc., to perform a system-level analysis of an ICE. Figure 3.6 shows the layout of the GT-Suite model of a 4-valve single-cylinder heavy-duty diesel engine built using measured dimensions, bounded at the intake and exhaust sides at the pressure measurement locations in the experimental engine.

GT-Suite discretizes the pipes and volumes connecting the cylinder to several sub-volumes of specified dimensions coupled one-dimensionally in the flow direction. Ordinary differential equations for the conservation of mass, momentum, and energy as written in equations (3.6), (3.7), and (3.8), respectively, are solved in each discretized volume [9]:

$$\frac{dm}{dt} = \sum_{\text{boundaries}} \dot{m} \tag{3.6}$$

$$\frac{d(me)}{dt} = -\rho \frac{dV}{dt} + \sum_{\text{boundaries}} (\dot{m}H) - hA_s (T_{\text{fluid}} - T_{\text{wall}}) \tag{3.7}$$

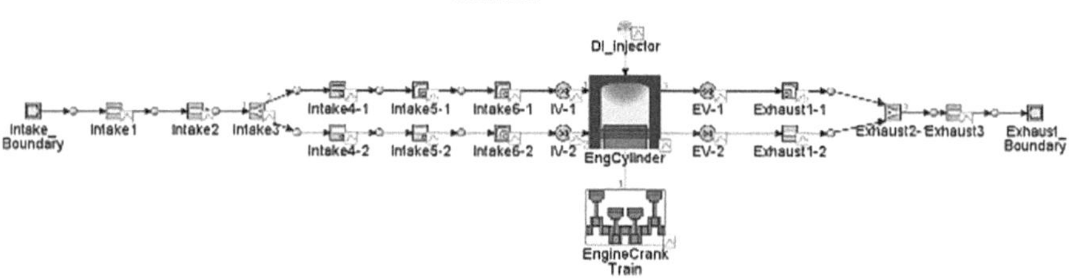

FIGURE 3.6 GT-Suite model layout of the heavy-duty single-cylinder diesel engine.

$$\frac{d\dot{m}}{dt} = \frac{-dpA + \sum_{\text{boundaries}} (\dot{m}u) - 4C_f \frac{\rho u|u|}{2} \frac{dxA}{D} - K_p \left(\frac{1}{2} \rho u|u|\right) A}{dx} \tag{3.8}$$

m – mass; $\dot{m}$ – mass flux into volume; e – total specific internal energy; ρ – fluid density; V – volume; H – total specific enthalpy; A_s – heat transfer surface area; h – convective heat transfer coefficient; T_{fluid} – fluid temperature; T_{wall} – wall temperature; p – pressure; u – velocity at the boundary; $|u|$ – mean velocity in the volume; A – cross-sectional flow area; C_f – Fanning friction factor; K_p – pressure loss coefficient

These equations are solved and integrated at each sub-volume and time step for calculating variables such as mass, velocity, pressure, temperature, and enthalpy. The integration is performed using an explicit method with a time step restricted by the Courant condition $\left(\frac{\Delta t}{\Delta x}(|u| + c) \leq 0.8\right)$. Since the pipes and sub-volumes are discretized only in the flow direction, friction loss and pressure drop due to geometry are accounted for in the equations using coefficients C_f and K_p, respectively. These coefficients can be calibrated at each individual pipe or volume to capture the manifold dynamics accurately.

This section details some of the significant sub-models of the multi-physics simulation tool, such as the valve, combustion, and thermal. The application of these sub-models to diesel (compression ignition) and gasoline (spark ignition) engines has been demonstrated. Some of the equations from the GT-Suite manual are used in this section to provide background for discussing the analyses. More detailed information about the models can be found in the GT-Suite manuals for flow modeling theory and engine performance.

3.3.1 Valve Model

A salient phenomenon of an ICE that requires careful examination is the pressure drop or mass flow rate across the valves which is dictated by equation (3.9). Since the geometrical effects of a valve on pressure drop or mass flow rates cannot be captured in the 1-D flow simulation, discharge coefficients (C_D) as a function of valve lift are used to approximate the losses [9]:

$$\dot{m} = C_D A_R \rho_{is} U_{is} \tag{3.9}$$

$\dot{m}$ – mass flow rate; A_R – reference flow area; ρ_{is} – density at the throat; U_{is} – isentropic velocity at the throat

Discharge coefficients must be determined from measured data to capture mass flow rates across the valves accurately. Figure 3.7 shows the discharge coefficients for intake and exhaust valves calculated using the pressures and mass flow rates from a three-dimensional computational fluid dynamics (3-D CFD) simulation for the same heavy-duty diesel engine discussed above at rated power condition. Figure 3.8 shows that the mass flow rates predicted by GT-Suite using the calculated discharge coefficients across the intake and exhaust valves closely matching the 3-D CFD.

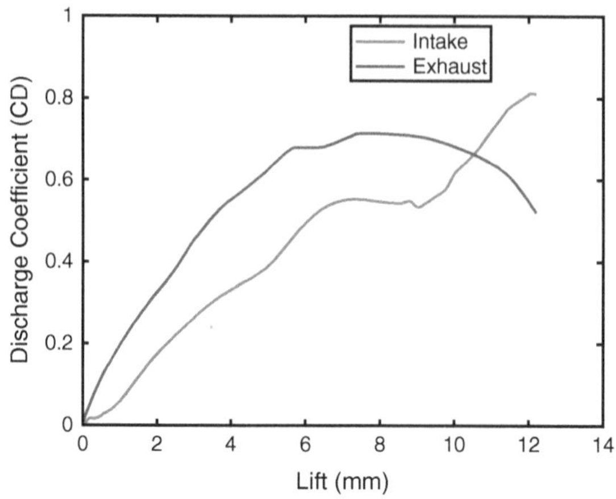

FIGURE 3.7 Discharge coefficients for intake and exhaust valves calculated using data from 3-D CFD simulation of a heavy-duty diesel engine at rated power condition.

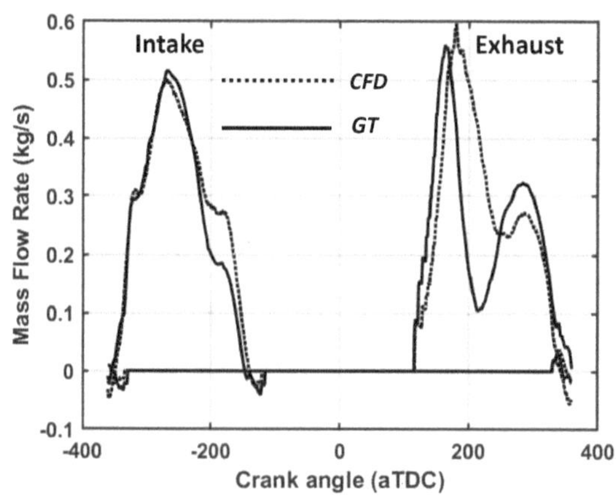

FIGURE 3.8 Mass flow rates across the intake and exhaust valves of a heavy-duty diesel engine predicted by GT-Suite and 3-D CFD for the rated power.

3.3.2 Combustion Model

GT-Suite uses a zero-dimensional approach for modeling combustion inside the cylinder, meaning that the entire cylinder is treated as one zone or region. Based on the model's predictive capability, the GT-Suite combustion models are classified as non-predictive, semi-predictive, and predictive combustion models.

3.3.2.1 Non-predictive

Non-predictive models are used where the combustion sensitivities are ignored. The burn rate, i.e., the fuel mass burned per unit time, is provided as the input for a non-predictive combustion model. The burn rate can be calculated from the cylinder pressure using the three-pressure analysis (TPA) method in GT-Suite, which requires measured cylinder, intake, and exhaust pressure data. Also, the fuel injection rate, the mass flows, and the emissions data are required for the TPA analysis. Figure 3.9 shows the burn rate calculated for the rated power condition of the single-cylinder heavy-duty diesel engine using the TPA method. Figure 3.10 shows the cylinder pressure predicted by the non-predictive combustion model using the calculated burn rate compared to the measured cylinder pressure data. Some discrepancies are expected due to in-cylinder flow assumptions made during the burn rate calculations. Wiebe functions, which could approximate combustion burn rates, can also be used to impose the burn rates without requiring the TPA calculation process.

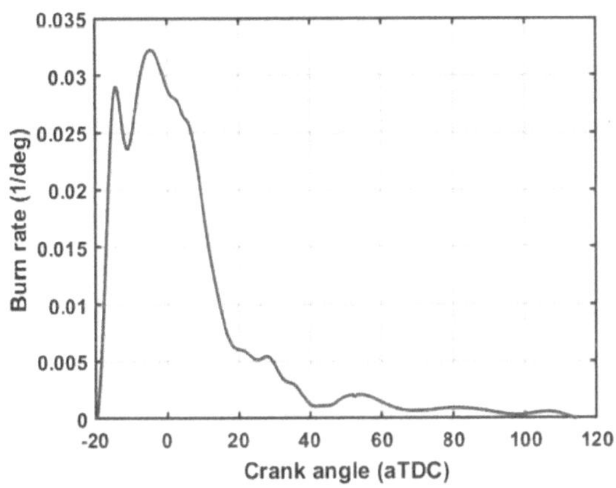

FIGURE 3.9 Burn rate calculated using the GT-Suite TPA analysis for the rated power.

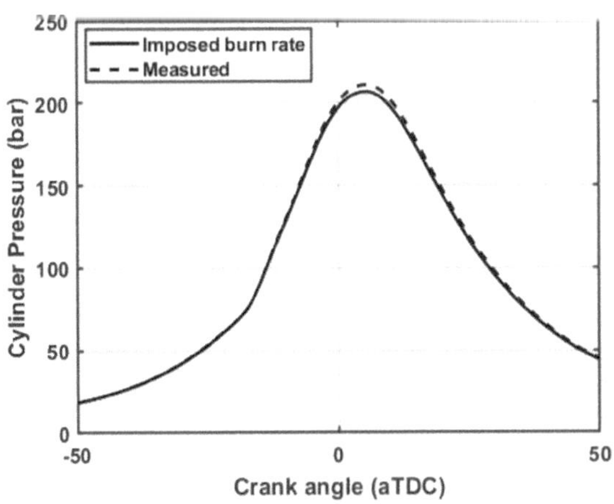

FIGURE 3.10 Comparison of cylinder pressures predicted with GT-Suite non-predictive combustion model and measurement for rated power SOI 29° bTDC.

3.3.2.2 Semi-predictive

Semi-predictive combustion models are similar to non-predictive combustion models, where the imposed burn rates or Wiebe functions can be provided as a look-up table based on parameters such as ignition timing, speed, and load, which could mimic a predictive combustion model.

3.3.2.3 Predictive

Predictive combustion models are used for GT-Suite studies that aim to analyze combustion and the factors affecting combustion. The in-cylinder environment prior to combustion is calculated, accounting for the manifold and valve flow dynamics, and the thermodynamics of the compression process. There are several models available in GT-Suite for predicting combustion.

3.3.2.3.1 Compression Ignition (CI) Combustion

'DI Pulse' is a commonly used combustion model in GT-Suite for diesel combustion. Since diesel combustion is very sensitive to the injection rate profile, the measured injection rate is provided as the input. From the fuel injection rate and nozzle geometry parameters, the following parameters are calculated (equations 3.10–3.17) consequently to arrive at the burn rate.

Injection velocity:

$$u_{\text{inj}} = C_d \frac{\dot{m}_{\text{inj}}}{A_n \rho_l} \tag{3.10}$$

Break-up time:

$$t_b = 4.351 \sqrt{\frac{2\rho_l}{\rho_g}} \frac{d_n}{C_d u_{\text{inj}}} \tag{3.11}$$

Spray tip length:

$$S = \left\{ u_{\text{inj}} t \left[1 - \frac{1}{16} \left(\frac{t}{t_b} \right)^8 \right] \frac{t}{t_b} \leq 1 \; u_{\text{inj}} t_b \frac{15}{16} \left(\frac{t}{t_b} \right)^{0.5} \frac{t}{t_b} \geq 1 \right\} \tag{3.12}$$

Velocity at spray tip:

$$u = \frac{dS}{dt} \tag{3.13}$$

Entrainment rate:

$$\frac{dm}{dt} = -C_{\text{ent}} \frac{m_{\text{inj}} u_{\text{inj}}}{u^2} \frac{du}{dt} \tag{3.14}$$

Ignition delay:

$$\tau_{\text{ign}} = C_{\text{ign}} \rho^{-1.5} e^{3500/T} \left[O_2 \right]^{-0.5} \tag{3.15}$$

Premixed burn rate:

$$\frac{dm_{\text{pm}}}{dt} = C_{\text{pm}} m_{\text{pm}} k \left(t - t_{\text{ign}} \right)^2 f([O_2]) \tag{3.16}$$

Diffusion burn rate:

$$\frac{dm}{dt} = C_{\text{df}} mk \frac{\sqrt{k}}{\sqrt{V_{\text{cyl}}}} f([O_2]) \tag{3.17}$$

C_d – nozzle discharge coefficient; $\dot{m}_{\text{inj}}$ – injection mass flow rate; A_n – injector nozzle area; ρ_l – liquid fuel density; ρ_g – gaseous fluid density; d_n – nozzle diameter; t – time; u – velocity at spray tip; $[O_2]$ – oxygen concentration; T – pulse temperature; ρ – pulse gas density; k – turbulent kinetic energy; t_{ign} – time at ignition; V_{cyl} – cylinder volume

3.3.2.3.2 Calibration of Combustion Constants

The predictive combustion models use tuning constants to calibrate the model for the engine and operating conditions in consideration. The DI Pulse combustion model has four combustion constants including an entrainment rate multiplier (C_{ent}), ignition delay multiplier (C_{ign}), premixed burn rate multiplier (C_{pm}), and diffusion burn rate multiplier (C_{df}) seen in above equations. The integrated design optimizer (IDO) tool in GT-Suite can be used to optimize the combustion constants to match experimental data. The burn rate calculated using the measured pressure analysis is used to compare the burn rates predicted for combustion constant sweeps by the IDO to find the optimum combination of constants that minimizes error.

Figure 3.11 shows the RMS error calculated by comparing the experimental net heat release rates of the single-cylinder heavy-duty diesel engine at rated power for

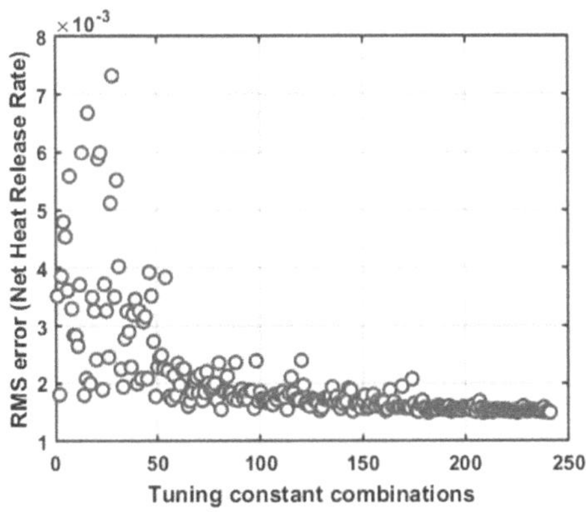

FIGURE 3.11 Net heat release rate RMS errors using 241 combinations of DI Pulse combustion constants predicted by IDO.

241 combinations of combustion constants. Figure 3.12 shows the predicted cylinder pressure by the DI Pusle model with the optimum combustion constants and the measurements for the injection timing sweep at rated power. After optimizing the combustion constants, an excellent agreement between measurement and predictions can be seen across the injection timing sweep.

3.3.2.3.3 Spark-Ignition (SI) Combustion

Spark-ignition combustion is characterized by the chemical kinetics and turbulence that dictate flame propagation, and 'SI Turb' is the commonly used combustion model for SI combustion in GT-Suite. The SI Turb combustion model considers the cylinder geometry, in-cylinder flow, thermodynamic environment, spark timing, and fuel properties. The burn rate calculation starts with the calculation of laminar and turbulent flame speeds, followed by the entrainment rate, and burn rate calculations (equations 3.18–3.22) [9].

Laminar flame speed:

$$S_L = (B_m + B_\varphi(\varphi - \varphi_m)^2\left(\frac{T_u}{T_{\text{ref}}}\right)^\alpha \left(\frac{p}{p_{\text{ref}}}\right)^\beta (1 - 0.75C_{\text{DE}}\left(1 - (1 - 0.75\ RGF)^7\right) \quad (3.18)$$

Turbulent flame speed:

$$S_T = C_{\text{TFS}}u'\left(1 - \frac{1}{1 + C_{\text{FKG}}\left(\frac{R_f}{L_i}\right)^2}\right) \quad (3.19)$$

Time constant:

$$\tau = \frac{C_{\text{TLS}}L_i}{S_L\sqrt{\text{Re}_t}} \quad (3.20)$$

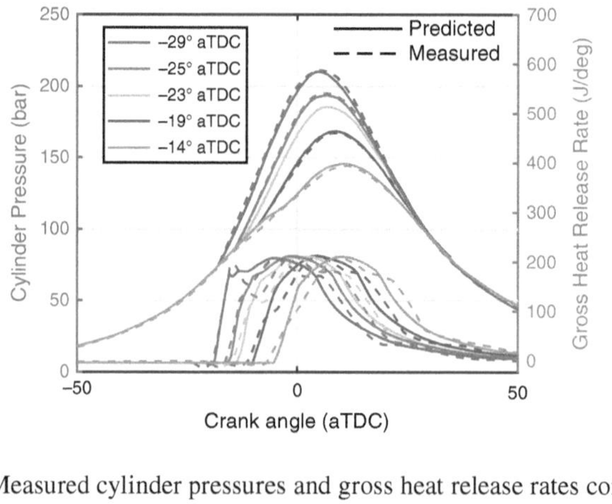

FIGURE 3.12 Measured cylinder pressures and gross heat release rates compared with the predictions by GT-Suite.

Entrainment rate:

$$\frac{dm_e}{dt} = \rho_u A_e \left(S_T + S_L\right) \tag{3.21}$$

$$\text{Burn rate} = \frac{dm_b}{dt} = \frac{m_e - m_b}{\tau} \tag{3.22}$$

B_m – maximum laminar speed; B_φ – laminar speed roll-off value; φ – equivalence ratio; φ_m – equivalence ratio at maximum speed; T_u – unburned gas temperature; T_{ref} – reference temperature; p_{ref} – reference pressure; p – pressure; α – temperature exponent; β – pressure exponent; RGF – residual gas fraction; u' - turbulence intensity; R_f – flame radius; L_i – integral length scale; Re_t – turbulent Reynolds number; ρ_u – unburned density; A_e – surface area at flame front

Similar to the DI Pulse model, the SI Turb model has four combustion constants to tune such as the dilution effect multiplier (C_{DE}), flame kernel growth multiplier (C_{FKG}), turbulent flame speed multiplier (C_{TFS}), and Taylor length scale (C_{TLS}) multiplier. Figure 3.13 shows the GT-Suite layout of a single-cylinder SI engine and Figure 3.14 shows the comparison of measured and predicted cylinder pressures using the SI Turb model with calibrated combustion constants for four different operating conditions listed in Table 3.1.

3.3.3 Thermal Model

GT-Suite solves for the temperature of each layer of a pipe sub-volume thickness by solving the energy conservation equation, which yields equation (3.23) after integration, where the change in internal energy of a solid pipe sub-volume layer is equal to the sum of heat fluxes across each boundary of the sub-volume:

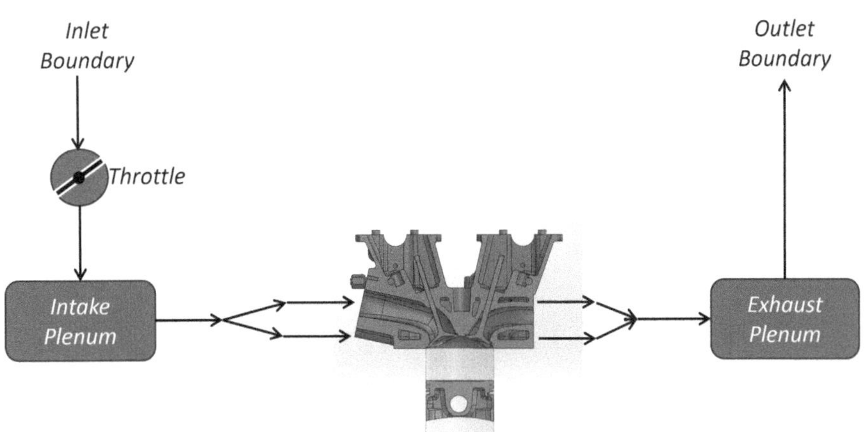

FIGURE 3.13 GT-Suite layout of a single-cylinder DISI engine.

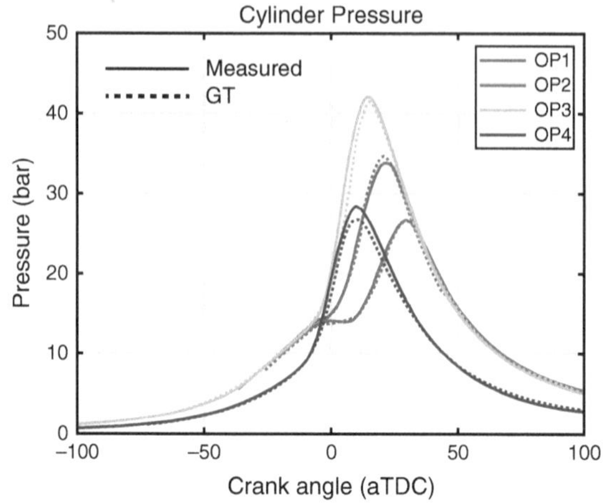

FIGURE 3.14 Measured and predicted cylinder pressures using the 'SI Turb' combustion model for four operating conditions.

TABLE 3.1
Operating Conditions Used for Calibrating 'SI Turb' Combustion Constants

	OP1	OP2	OP3	OP4
RPM	1200	1500	1800	1500
Spark timing (CAD aTDC)	−6.9	−15.4	−21.1	−32.1
Net IMEP (bar)	7.04	7.56	8.1	4.07

$$\rho C_v \frac{\Delta T}{\Delta t} = \sum_{\text{faces}} -qA \tag{3.23}$$

ρ – density; C_v – specific heat; ΔT – change in temperature; Δt – change in time; q – heat flux; A – face surface area

The heat fluxes across each face or boundary are calculated using the standard equations for conduction, radiation, and convection (see equations (3.24), (3.25), and (3.26) respectively).

Conduction:

$$q = -k.\nabla T \tag{3.24}$$

Radiation:

$$q = -\varepsilon\sigma\left(T_1^4 - T_2^4\right) \tag{3.25}$$

Convection:

$$q = h\left(T_g - T_w\right) \tag{3.26}$$

k – thermal conductivity; ε – emissivity; σ – Stefan-Boltzmann constant; T – surface temperature; h – convective heat transfer coefficient; T_g – gas temperature; T_w– wall temperature

Figure 3.15 shows the internal wall temperatures of the intake and exhaust pipes of the single-cylinder heavy-duty diesel engine predicted by the steady-state thermal solver in GT-Suite for rated power condition. The internal wall temperature profile as a function of pipe radius of an exhaust pipe is plotted in Figure 3.16 where 0 mm corresponds to the internal surface temperature. The internal surfaces exchange heat with the working gas whose heat transfer coefficients are calculated using heat transfer correlations. Colburn's heat transfer analogy [10] is a commonly used correlation for turbulent flow through pipes shown in equation (3.27). The external surfaces are

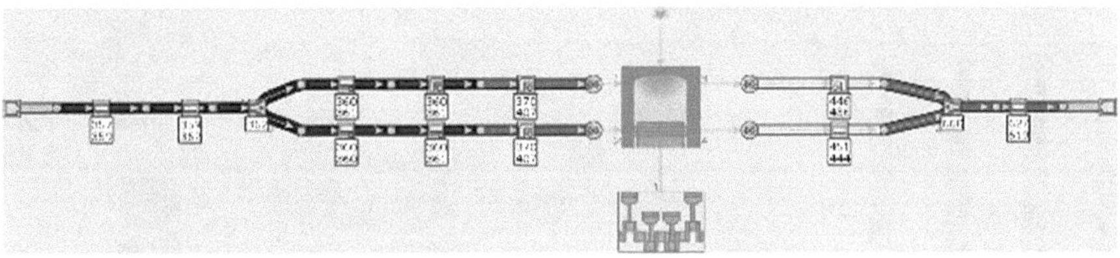

FIGURE 3.15 Internal wall temperatures of each pipe connected to the cylinder of the heavy-duty diesel engine at rated power predicted by GT-Suite thermal solver.

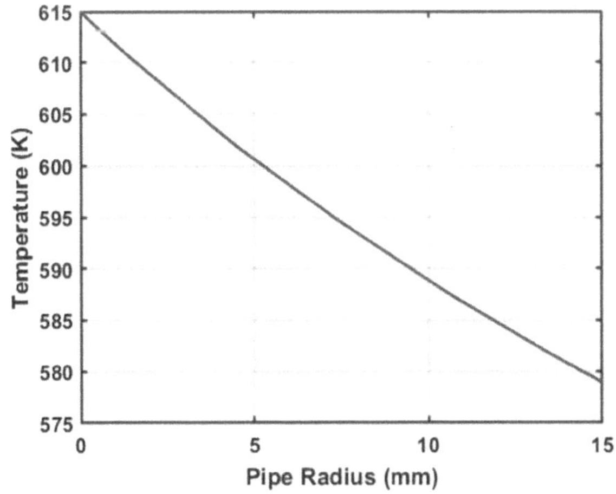

FIGURE 3.16 Internal wall temperature of an exhaust pipe of the heavy-duty diesel engine as a function of radius predicted by GT-Suite thermal solver.

cooled by either ambient air or coolant whose heat transfer coefficients are usually assumed and calibrated based on measurements:

$$h = 0.5 C_f \rho U_{\text{eff}} C_p \text{Pr}^{-2/3} \tag{3.27}$$

C_f – Fanning friction factor; ρ – density; U_{eff} – effective velocity; C_p – specific heat; Pr – Prandtl number

The combustion chamber wall temperatures are predicted using a coarse finite element thermal analysis of a geometrical structure modeled based on measured dimensions. The energy conservation equation is solved at each finite volume. The predicted temperatures of the heavy-duty diesel engine combustion chamber components at rated power are shown in Figure 3.17. The combustion side convective heat transfer coefficients are calculated using the Woschni heat transfer model [3] and the oil/coolant heat transfer coefficients are assumed. Following is an example case study involving the application of thermal barrier coatings (TBCs) on combustion chamber surfaces performed utilizing the GT-Suite thermal models.

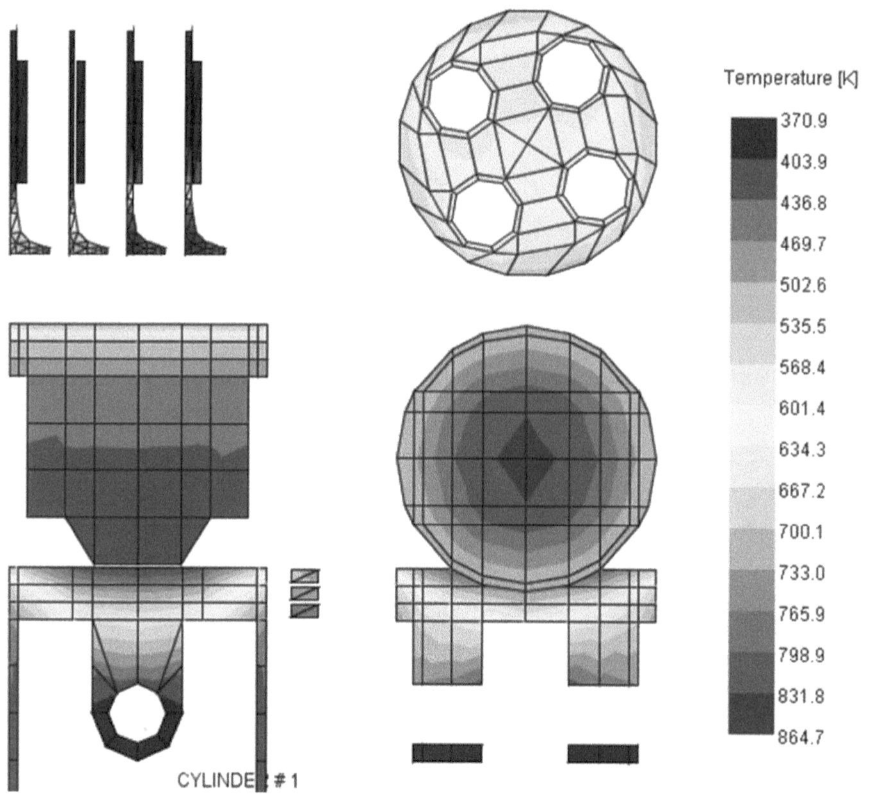

FIGURE 3.17 2-D views of the combustion chamber component temperatures predicted by GT-Suite coarse 3-D finite element thermal analysis.

3.3.4 Application of Thermal Barrier Coatings

Thermal barrier coatings (TBCs) are popularly used in gas turbines to block heat transfer and protect the turbine blades. These coatings with low thermal inertia could swing their surface temperatures in phase with the gas temperatures more effectively than aluminum or steel that have a higher thermal inertia, thereby reducing the temperature difference between the gas and the wall and, consequently, the convective heat transfer. TBCs are currently being investigated in ICEs for their potential thermal efficiency benefits. The commonly explored application of TBCs in ICEs is on the surfaces of the combustion chamber to block heat transfer from combustion [11].

Determining the surface temperature swing produced by these coatings is essential for the investigation of TBCs. However, the measurement techniques (commonly thermocouples) measure inaccurate values due to disturbances caused by the differences in thermophysical properties of the measuring probe and the coating substrate. Thus, mathematical modeling approaches are leveraged to predict surface temperature swings of the TBCs. GT-Suite can predict the temperature swing of the combustion chamber surfaces with the application of TBCs.

Figure 3.18 shows the predicted combustion chamber surface temperatures of a single-cylinder direct-injection spark-ignition (DISI) engine operated at 1500 RPM, 8 bar IMEP with and without TBCs. Material properties of the gadolinium zirconate (GdZr) TBC, a commercially available TBC, were used for the predictions. The TBC-coated surface produced an average temperature swing of 100 K, as predicted by the GT-Suite model, compared to the metal baseline (aluminum), whose temperature swing is negligible (~ 3 K). The liner surface was decided to be uncoated due to durability issues as the piston moves over the liner. The valves show relatively high temperatures since they are not cooled like the other combustion chamber surfaces.

Multi-physics simulation tools combining the several sub-models discussed above are valuable for system-level studies. To demonstrate the capability of GT-Suite for performing system-level studies, the following two case studies are used. The first

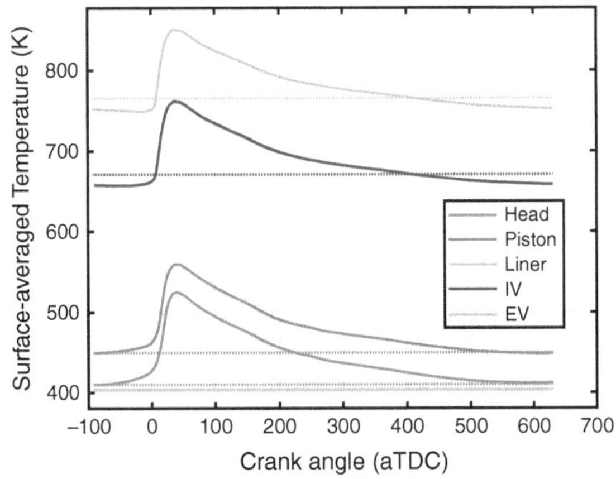

FIGURE 3.18 Surface temperatures of combustion chamber components predicted by GT-Suite with and without TBCs in a single-cylinder DISI engine.

case study is a system-level thermal analysis of a multi-cylinder SI engine for predicting the potential improvement of catalyst warmup rates by coating different surfaces of the engine. The second study shows the potential of GT-Suite for manifold dynamics tuning in a variable valve timing engine to maximize volumetric efficiency.

3.3.5 System-Level Effects of TBCs

There are some less explored application areas on an ICE where the TBCs could have potential benefits on a system level. One such potential application is the surfaces of the exhaust flow components, such as the ports, manifolds, runners, and pipes. Three-way catalytic converters (TWCs), the most common emission after-treatment system in modern SI engines, suffer from low emissions conversion efficiencies at low temperatures. During cold starts, the DISI engines emit a lot of unburned hydrocarbon and CO emissions due to poor combustion efficiency and colder walls [12]. Simultaneously, the low emissions conversion efficiencies of TWCs at colder conditions exacerbate the problem [13]. The large thermal inertia associated with the exhaust flow components and the turbocharger casing in a turbocharged engine absorbs exhaust enthalpy from the burnt gases, reducing the heating rate of the catalyst surfaces during cold starts.

TBCs, when coated on the combustion chamber surfaces and exhaust flow path (ports, manifolds, runners), could help increase the catalyst surface temperatures faster by blocking heat transfer to metal components and saving enthalpy for the catalyst during cold starts. A multi-cylinder turbocharged DISI engine was modeled in GT-Suite as shown in Figure 3.19, and the prospective use of TBCs on various surfaces of an ICE for faster catalyst warmup was explored [14].

Measured dimensions from the experimental engine were used to model the geometry in GT-Suite, and the model was calibrated with measured data. Figure 3.20 shows the predicted transient turbine inlet temperatures compared to measured slow-response thermocouple temperatures at the same location in the experimental engine for three low-load (2.5 bar IMEP), low-speed operating conditions.

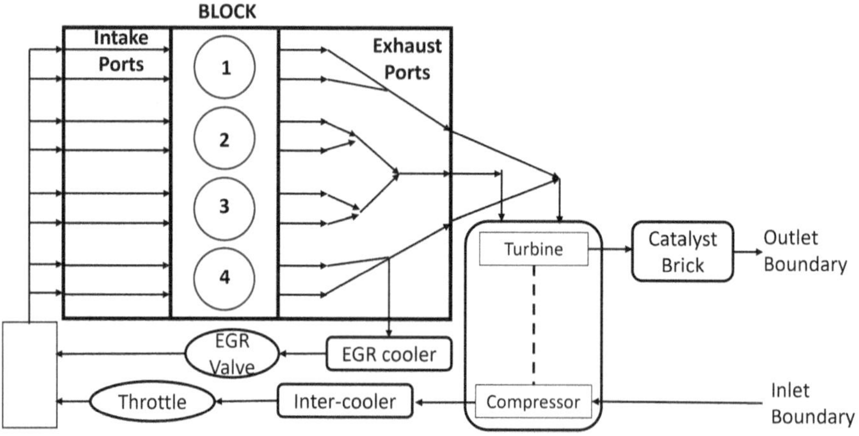

FIGURE 3.19 GT-Suite model layout of a multi-cylinder turbocharged DISI engine.

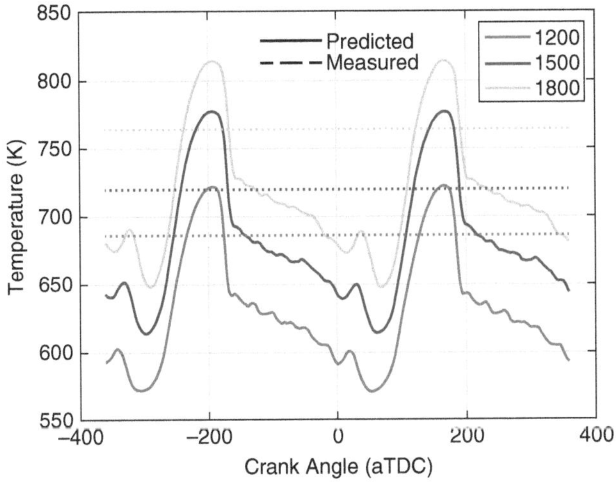

FIGURE 3.20 Predicted transient turbine inlet temperatures compared with the slow-response thermocouple temperatures for the three low-speed, low-load operating conditions.

The model was run at a typical cold-start operating condition (1600 RPM, 2 bar IMEP) for 45 seconds after motoring for initial 35 cycles to develop the flow. A GdZr coating was tested on different engine surfaces to study the effects on catalyst surface temperature. A transient thermal solver in GT-Suite was used to predict the surface temperature changes during the cold-start. Figure 3.21 shows the evolution of catalyst surface temperature during the 45 seconds of the cold-start for different iterations of TBC application areas where the dotted lines correspond to the time at which the catalyst surface temperatures reach 623 K. This temperature was assumed as the catalyst light off temperature (90% emissions conversion efficiency), and the time to reach the light-off temperature is usually defined as the catalyst light-off delay. The baseline is when no TBCs were applied on any of the surfaces.

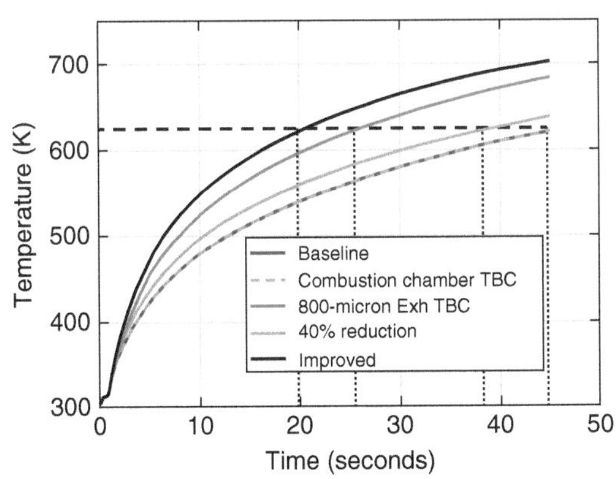

FIGURE 3.21 Catalyst surface temperatures as a function of time during cold-start operation predicted by GT-Suite model for TBCs on different surfaces of the multi-cylinder DISI engine.

When 120-micron GdZr coating was applied on all the combustion chamber surfaces, a negligible effect on catalyst surface temperature was predicted since the uncoated exhaust flow components absorbed the blocked heat transfer from the combustion chamber. When an 800-micron GdZr coating was applied on all the surfaces upstream of the catalyst and downstream of the cylinder where the exhaust gases exchange heat (except the turbocharger casing), a 20-second reduction in catalyst light-off delay was predicted, which could help reduce cold-start emissions. When the turbocharger casing thermal inertia was theoretically reduced by 40% since the feasibility of TBCs on turbocharger casings was not investigated before, the catalyst light-off delay improved by 8 seconds. The improved curve in the plot refers to the iteration where the combined benefits of the three approaches discussed above were tested. A 25-second reduction (more than 50% reduction from the baseline) in the catalyst light-off delay was predicted, which could significantly reduce tailpipe emissions during cold starts.

3.3.5.1 Manifold Dynamics

Manifold dynamics are another important system-level aspect commonly studied using multi-physics simulation tools. In a multi-cylinder engine, intake and exhaust manifold dynamics are crucial for volumetric efficiency. Thus, modern engines use variable valve timing to maximize the volumetric efficiency. Figures 3.22 and 3.23 show the pressure fluctuations in the intake and exhaust manifolds, respectively, during a motoring cycle of a turbocharged DSI engine predicted by GT-Suite motored at four different speeds. The manifold dynamics strongly correlate with engine speed, and the valve timings are tuned to maximize volumetric efficiency at each engine speed. GT-Suite can be leveraged to tune for the optimum valve timings at each speed. Figure 3.24 shows the mass flow rates predicted by GT-Suite after tuning the valve timings individually for each engine speed compared to the measurement.

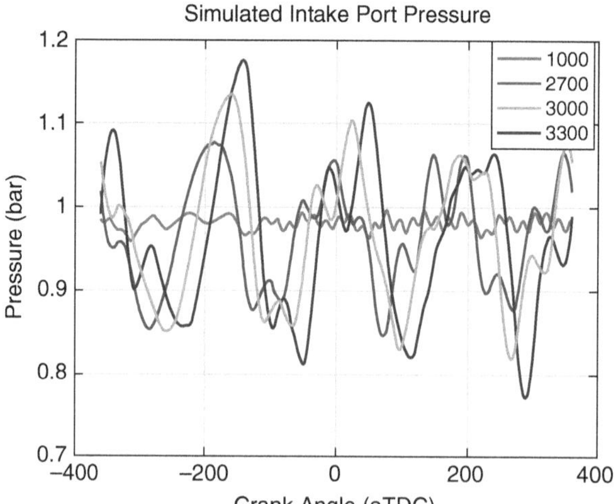

FIGURE 3.22 Intake manifold pressures predicted by GT-Suite for the multi-cylinder DISI engine motored at four different speeds.

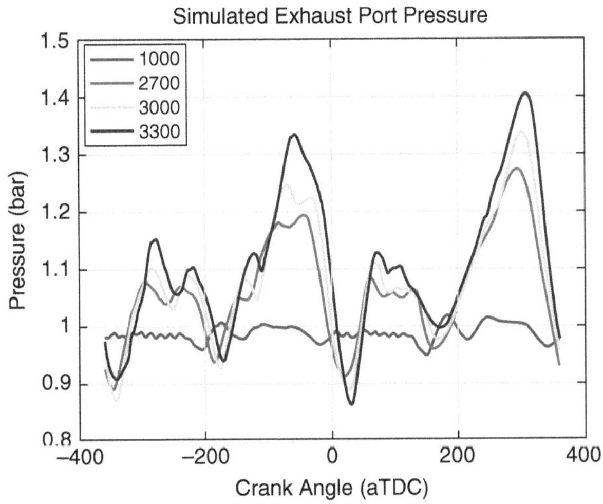

FIGURE 3.23 Exhaust manifold pressures predicted by GT-Suite for the multi-cylinder DISI engine motored at four different speeds.

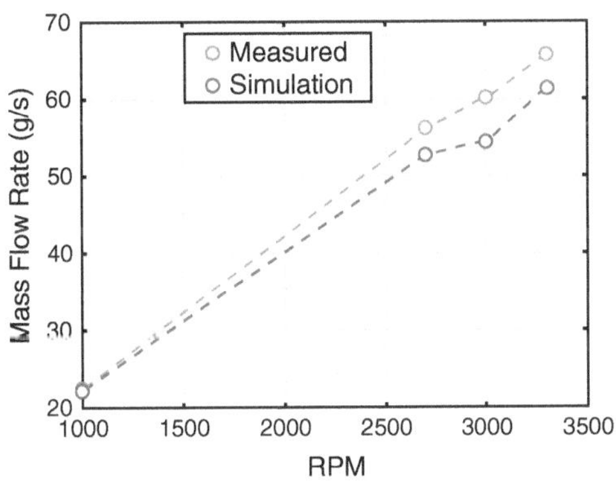

FIGURE 3.24 Mass flow rates predicted by GT-Suite after valve timing tuning compared with measured data.

GT-Suite is often used as a tool to study the sensitivity of intake and exhaust flow pathways for modifying the manifold dynamics.

The demonstrated system-level simulation case studies show the relevance of multi-physics simulations with reduced-order models for ICE research. The forte of this tool is the capability to perform system-level studies at affordable computational expense. A detailed discussion about the computational expense will be provided later. Moreover, the competence of the tool to capture a certain degree of spatial stratification increases the usefulness of the reduced-order models. Also, the amount of experimental data needed for a reduced-order model is greatly reduced since ambient-to-ambient simulations are possible. Nonetheless, the sub-models require experimental or high-fidelity simulation data for tuning, which is typical of reduced-order models. This can be seen from the valve model discharge coefficient calculation, combustion model constant tuning, thermal model cooling boundary condition tuning, etc., which were discussed previously. To reduce the assumptions and tuning, high-fidelity three-dimensional simulations are performed on ICEs for more accurate solutions, traded for increased computational expense.

3.4 SECTION-3: HIGH-FIDELITY SIMULATIONS

High-performance computing has enabled the ICE research community to perform simulations that can almost mimic all of the physical phenomena in an ICE to the finest scale using high-fidelity multi-dimensional mathematical models. The underlying principle is that any physical control volume can be discretized into small enough volumes, and the governing equations can be solved digitally in each volume and among volumes to track the flow of energy or mass. Some valuable high-fidelity simulation tools used for ICE research such as three-dimensional computational fluid dynamics (3-D CFD) and finite element analysis (FEA) models are discussed below, with case studies illustrating their application to a heavy-duty diesel engine. Furthermore, this section contains discussion about the sub-models and boundary conditions essential for the high-fidelity simulations. A co-simulation methodology that combines the two high-fidelity numerical tools used for heat transfer studies is detailed following the CFD and FEA model descriptions. Results from an injection parameter sensitivity study using the co-simulation methodology on a heavy-duty diesel engine piston are then discussed. Converge CFD and Abaqus are the software tools used for the analyses discussed in this section. Some equations from the respective manuals of the software tools are used to explain the modeling theory.

3.4.1 3-D CFD

Since the dominant working substances in ICEs are fluids, 3-D CFD can be used to predict the fluid flow in an ICE. The flow prediction combined with chemistry solvers can predict the combustion. The governing equations for predicting 3-D fluid flow in CFD are conservation/transport equations for the mass, momentum, and energy, commonly called as Navier-Stokes equations, shown in equations (3.28), (3.29), and (3.30), respectively [15]:

$$\frac{\partial \rho}{\partial t}+\frac{\partial \rho u_i}{\partial x_i}=S \tag{3.28}$$

$$\frac{\partial \rho u_i}{\partial t}+\frac{\partial \rho u_i u_j}{\partial x_j}=-\frac{\partial P}{\partial x}+\frac{\partial \sigma_{ij}}{\partial x_j}+S_i \tag{3.29}$$

$$\frac{\partial \rho e}{\partial t}+\frac{\partial \rho e u_j}{\partial x_j}=-P\frac{\partial u_j}{\partial x_j}+\frac{\partial}{\partial x_j}\left(K\frac{\partial T}{\partial x_j}\right)+\sigma_{ij}\frac{\partial u_j}{\partial x_j}+\frac{\partial}{\partial x_j}\left(\rho\sum_m D_m h_m\frac{\partial Y_m}{\partial x_j}\right)+S \tag{3.30}$$

u_i, u_j – velocities in directions i, j; ρ – density; P – pressure; σ_{ij} – viscous stress tensor; e – internal energy; K – thermal conductivity; T – temperature; D_m – species mass diffusion coefficient; Y_m – species mass fraction; h_m – species-specific enthalpy; ∂t – change in time; ∂x – change in space

As discussed in the 1-D heat conduction model, these continuous equations can be discretized using various numerical schemes. 3-D CFD discretizes the control volume into multiple volumes/cells based on the chosen grid size, and the discretized transport equations can predict the flow of mass, momentum, and energy between the cells. The term S seen in the equations is the internal generation term. For example, the energy released from combustion is accounted for in each cell using the term S in the energy equation. Similarly, mass added or removed from a cell is accounted using the term S in the mass transport equation. For a stable solution, the time step should be chosen based on the grid size used. The Courant-Friedrichs-Lewy (CFL) number governs the proportionality between the grid size (Δx) and the time step (Δt) considering the maximum velocity (u). The CFL number given by equation (3.31) denotes the maximum number of cells a variable can travel in a time step:

$$u \cdot \frac{\Delta t}{\Delta x} < \text{CFL} \tag{3.31}$$

Even the three-dimensional high-fidelity models use sub-models or assumptions to efficiently compute the solution. The necessity and limits of sub-models used with 3-D CFD can be understood from the following sub-section that discusses turbulence and their relevance to ICE simulations. Other important sub-models used with 3-D CFD such as the spray sub-models and the wall heat transfer (near-wall) sub-models are also discussed.

3.4.1.1 Turbulence

The most challenging problem or physical fluid dynamics phenomenon to capture in the computational space is turbulence. Turbulence is the irregular motion of fluids frequently appearing as circular structures in the fluid called eddies influenced by multiple factors. The turbulent eddies promote the mixing of fuel and oxidizer inside the cylinder and, therefore, greatly influence combustion. Thus, accounting for the turbulence effects is necessary to predict combustion accurately. These eddies can be in various scales ranging from the entire size of a control volume to the finest scale in a control volume, the smallest of the eddies dissipating as heat due to viscous forces. In terms of an ICE, the most considerable eddy would be the size of the cylinder (i.e., the bore and/or stroke) and the smallest eddy could be microns.

Turbulent kinetic energy is the widely used terminology used to quantify the potential of the eddies for enhanced mixing. Figure 3.25 shows the turbulent kinetic energy as a function of the turbulent eddy length scales and the level of eddies captured by some popular CFD approaches to capture turbulence [16]. The large eddies contain the maximum kinetic energy, which is then transferred to the smaller eddies, and the smaller eddies finally dissipate the kinetic energy as heat due to viscous forces. The smallest length scale containing non-negligible kinetic energy in a control volume is given by the Kolmogorov length scale $\left(\eta = \frac{\upsilon^3}{\varepsilon}^{1/4}\right)$ determined

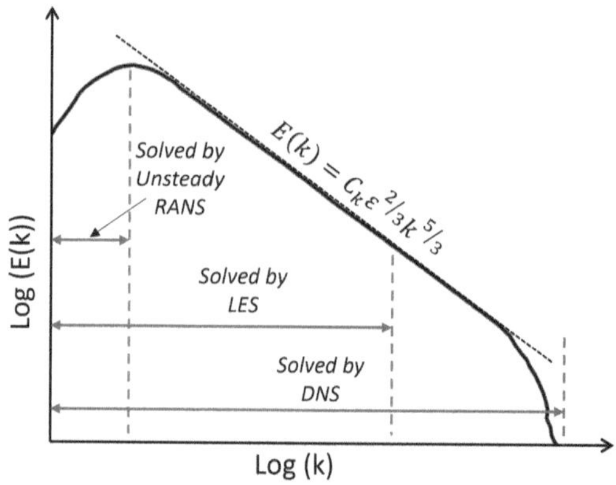

FIGURE 3.25 Turbulent kinetic energy as a function of turbulent eddy length scales and the length scales solved by each of the different CFD approaches to solve for turbulence [16].

using kinematic viscosity of the fluid (υ) and the dissipation rate (ε). Directly solving the Navier-stokes equations with a small enough grid size that could capture the Kolmogorov length scales is called direct numeric simulations (DNS) and is considered computationally impractical for an ICE simulation. Thus, alternate techniques have emerged to capture the turbulence effects with affordable computational expense based on the length scales of eddies solved.

3.4.1.1.1 Reynolds-Averaged Navier-Stokes (RANS)

RANS is a popular CFD technique where the mean flow is captured using a coarser grid, and the turbulence effects are accounted for using mathematical models. Since the objective of this type of simulation is to capture the mean flow, the velocity (u) is written as a combination of an ensemble-averaged term $(\underline{u})$ and a fluctuating term (u') (equation 3.32) [15].

$$u = \underline{u} + u' \tag{3.32}$$

Substituting this into the original Navier-Stokes equation provides the Reynolds-averaged Navier-Stokes (RANS) equations. For example, RANS momentum transport equation is shown in equation (3.33) [15]. Favre-average $\left(\tilde{u} = \frac{\underline{\rho u}}{\underline{\rho}}\right)$ is used for compressible flow to account for density variations.

$$\frac{\partial \underline{\rho} \tilde{u}_i}{\partial t} + \frac{\partial \underline{\rho} \tilde{u}_i \tilde{u}_j}{\partial x_j} = -\frac{\partial \underline{P}}{\partial x} + \frac{\partial}{\partial x_j}\left[\mu\left(\frac{\partial \tilde{u}_i}{\partial x_j} + \frac{\partial \tilde{u}_j}{\partial x_i}\right) - \frac{2}{3}\mu \frac{\partial \tilde{u}_k}{\partial x_k}\delta_{ij}\right] + \frac{\partial}{\partial x_j}\left(-\underline{\rho}\widetilde{u_i' u_j'}\right) \tag{3.33}$$

The final term in the right-hand side of the equation is known as the *Reynolds stress tensor* that accounts for the turbulence effects. The Reynolds stress tensor is modeled using mathematical models called as turbulence sub-models that account for the influence of turbulent eddies rather than solving them directly.

3.4.1.1.2 Large Eddy Simulations (LES)

Figure 3.25 shows that the larger length scale eddies contain the bulk of the kinetic energy compared to the smaller length scales. The large eddy simulations (LES) are used to solve for those energy-containing large eddies and mathematically model the kinetic energy available in the smaller eddies. The modeled part of the turbulence is called the *sub-grid scale* since the grid cannot resolve them. Similar to the RANS, the LES models the variables (for example, velocity (u)) as a combination of a spatially filtered term $(\langle u \rangle)$, which the grid resolves and a sub-grid term (u''), as shown in equation (3.34) [15]. The spatial filtering of a variable is like applying a box filter in space, implying that the variable accounts only for the eddies resolved using the grid. Applying this definition of velocity to the momentum conservation equation gives the spatially filtered form shown in equation (3.35) [15]. The sub-grid scale (SGS) velocity induces an additional stress tensor that is modeled using mathematical models. It should be noted that the sub-grid term in LES and the fluctuating term in RANS are different, and therefore, should be modeled in different ways. Thus, different turbulence sub-models are required to solve the SGS stress tenor in LES:

$$u = \langle u \rangle + u'' \tag{3.34}$$

$$\frac{\partial \langle \rho \rangle \langle u_i \rangle}{\partial t} + \frac{\partial \langle \rho \rangle \langle u_i \rangle \langle u_j \rangle}{\partial x_j} = -\frac{\partial \langle P \rangle}{\partial x} + \frac{\partial \langle \sigma_{ij} \rangle}{\partial x_j} + \frac{\partial \tau_{ij}}{\partial x_j} \tag{3.35}$$

3.4.1.2 Unsteady RANS (URANS) vs. LES

When the RANS equations are used to solve for unsteady flow like in an ICE, the simulations are termed as unsteady RANS (URANS). Figure 3.26 shows a slice of an 8th sector of a heavy-duty diesel engine simulated with 1 mm mesh using the generalized $k-\varepsilon$ RANS turbulence model and dynamic structure LES turbulence model. Circular structures in the fluid are clearly seen in the LES compared to the RANS, where the flow looks more dampened or smoothened. The reason being that the turbulence sub-models used for solving the Reynolds stress in the RANS induce artificial viscosity in the flow, restricting the smaller structures from forming since

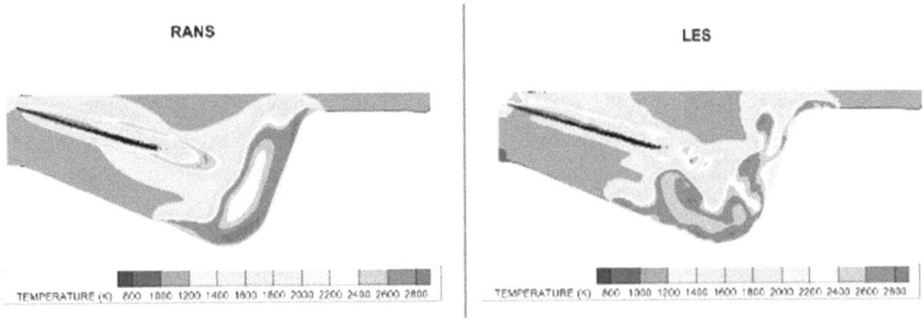

FIGURE 3.26 Temperature contour slice of a heavy-duty diesel engine simulated with RANS and LES turbulence models.

the objective is to capture the mean flow. On the other hand, the viscosity induced by the LES turbulence models is comparatively minor and has grid dependence, allowing the smaller structures to form depending on the grid size. With that said, an LES simulation with a small enough grid that could solve all the eddy length scales would make the SGS turbulent kinetic energy zero, which would be equivalent to DNS.

The fact that LES can capture a certain level of eddies dictated by the grid size makes it better suited than RANS to capture the cycle-to-cycle variations characteristic of ICEs. LES is considered more accurate than URANS, since it can capture realistic fluid flow behavior. However, LES simulations are computationally expensive than URANS due to the requirement of smaller grid sizes and the need for simulating more cycles than URANS if cycle-averaged results are desired. Nonetheless, URANS provides a computationally efficient solution where the small-scale perturbations of the system are not a concern.

A wide variety of analyses are discussed in the following sections, where the URANS simulations are used for computational efficiency. The URANS cannot converge to DNS even with a fine grid since the turbulent viscosity induced by the RANS turbulence models is independent of the grid, restricting the formation of smaller eddies. Thus, refining the grid beyond a certain level has insignificant effects on the results in a URANS simulation since there are no smaller flow structures to capture. However, some length scales still need to be resolved in URANS as seen in Figure 3.25 called the integral length scales, where the kinetic energy formation occurs. A grid convergence study is essential for URANS to find the largest grid that could capture all the integral length scales for a computationally efficient solution.

3.4.1.3 Grid Convergence Study

The grid resolution required for unsteady RANS (URANS) simulations can be determined by performing a grid convergence study where the grid will be refined until the changes are below a threshold [17]. Converge CFD is one of the well-known software packages available for performing CFD simulations in ICEs. Converge CFD is known for its innovative meshing techniques, which can ease the onerous task of manually meshing complex geometries like those found in modern ICEs. Adaptive mesh refinement (AMR) is a state-of-the-art mesh refinement technique used by Converge CFD to refine the mesh locally and dynamically based on the difference between the actual and grid-resolved values. AMR reduces the computational time significantly by allowing coarser grids in regions without significant gradients and refining the necessary regions. Additionally, the grid can be refined at specific regions during specific time periods of a simulation using ‘fixed embedding’.

Figure 3.27 shows a cut section of the heavy-duty diesel engine at TDC simulated using 3-D CFD-RANS technique for the rated power condition of the engine with a start of injection timing of 19° bTDC. A base grid of 1.4 mm was used throughout the cylinder. AMR based on velocity and temperature was set to capture the strong gradients induced by the diesel combustion and the liquid fuel spray dynamics, which can be seen from the refinements around the spray plumes. Also, the grid was refined near the valve seats to capture the pressure drop across the valves, which are visible in Figure 3.27. The smallest grid size used in the simulation was 0.35 mm, which is recommended for diesel engine simulations using RANS turbulence models

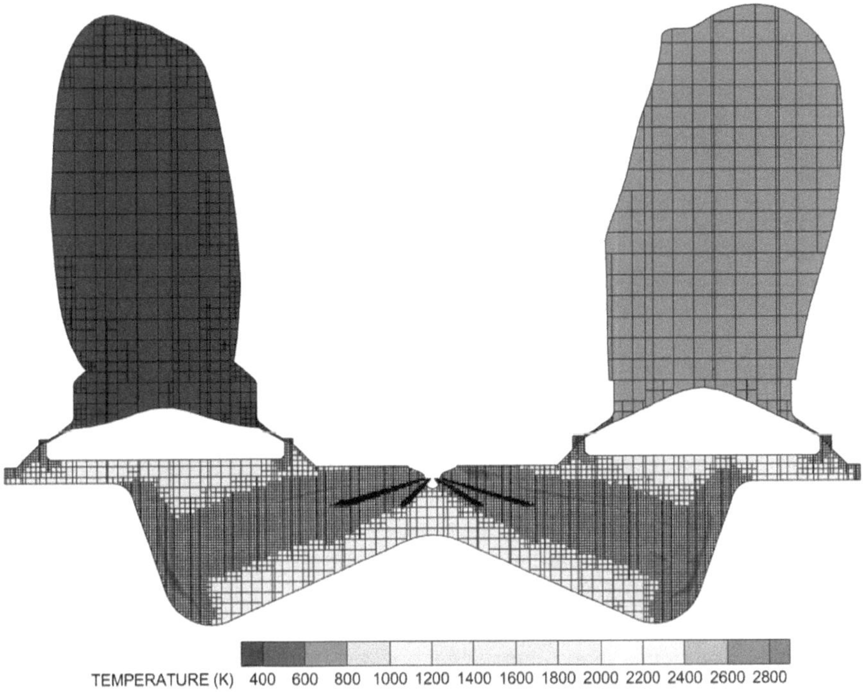

FIGURE 3.27 Cut section of the heavy-duty diesel engine CFD domain at combustion TDC showing the mesh and temperature contours predicted.

by Converge CFD. Without the AMR and embedding techniques, the entire CFD domain must be refined to 0.35 mm to capture the mean flow accurately, drastically increasing the computational expense. Renormalization group $k-\varepsilon$ (RNG $k-\varepsilon$), a commonly used turbulence sub-model, was used for this simulation.

To verify the largest grid size that could capture the integral length scales or mean flow, a grid convergence study for the RANS simulations of the diesel engine was conducted by changing the base grid with the mentioned embedding techniques, which scaled the local regions accordingly from the base grid. Figure 3.28 shows the cylinder pressures and gross heat release rates simulated by varying the base grid where the legend indicates the smallest grid size used in each case. The results significantly varied between 0.7 mm and 0.35 mm. However, the changes were minor between 0.35 and 0.175 mm, suggesting a grid size of 0.35 mm or smaller can be used for the simulation, which agrees with the converge CFD recommendation.

3.4.1.4 Spray Sub-Models

Spray dynamics are crucial for accurately predicting combustion in a diesel engine. Measured injection rate profile and mass are provided as inputs to the CFD for simulating the diesel combustion. Since fossil fuels such as diesel, gasoline, and jet fuel have multiple chemical species in them and significant property variance is expected, it is hard to determine all the species and their mass fractions. Thus, C14H30 (tetradecane), a well-established surrogate for the physical properties of diesel fuel,

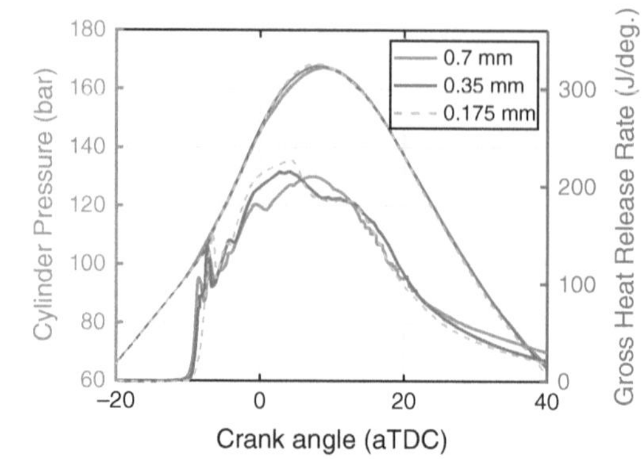

FIGURE 3.28 Cylinder pressures and gross heat release rates simulated by varying the base grid.

is used for the diesel engine simulations. The liquid fuel is injected as discrete parcels into the cylinder. The parcel sizes are based on the injector geometry, injection rate profile, and mass. The liquid parcels undergo several physical processes to evaporate, such as spray break-up, drop drag, collision and coalescence, and turbulent dispersion, as shown in Figure 3.29 [15]. The CFD simulations need sub-models to account for these different physical processes to determine the evaporation rate of each liquid parcel. Also, the liquid spray impingement on the combustion chamber walls and their evaporation require additional sub-models. The sub-models for capturing spray dynamics used for the diesel engine simulation are listed in the table below.

Figure 3.30 shows the liquid spray penetration predicted by the CFD-RANS for 2000 bar and 1000 bar injection pressures used for simulating the rated power condition with a start of injection timing of 29° bTDC in the heavy-duty diesel engine. The injection duration was increased by approximately 0.9 ms for the lower injection pressure. The higher injection pressure atomizes the fuel well, which increases the drag on individual droplets, thereby reducing the inertia of the fuel jet and consequently reducing the maximum liquid spray penetration length.

3.4.1.5 Near-Wall Model and Heat Transfer Sub-Model

Another challenging physical phenomenon to capture in the computational space is the boundary layer and wall heat transfer in ICEs. Due to the interaction between turbulence in the bulk gas and viscosity near the wall, steep velocity and temperature

TABLE 3.2
Spray Sub-models Used in the CFD Simulation of the Heavy-duty Diesel Engine

Spray Break-up	**Kelvin-Helmholtz and Rayleigh-Taylor [18]**
Evaporation	Frossling [19]
Turbulent Dispersion	O'Rourke [19]
Collision	No-time-counter [20]
Drop drag	Dynamic [21]
Spray-wall interaction	Wall film [22]
Wall splash	O'Rourke [22]

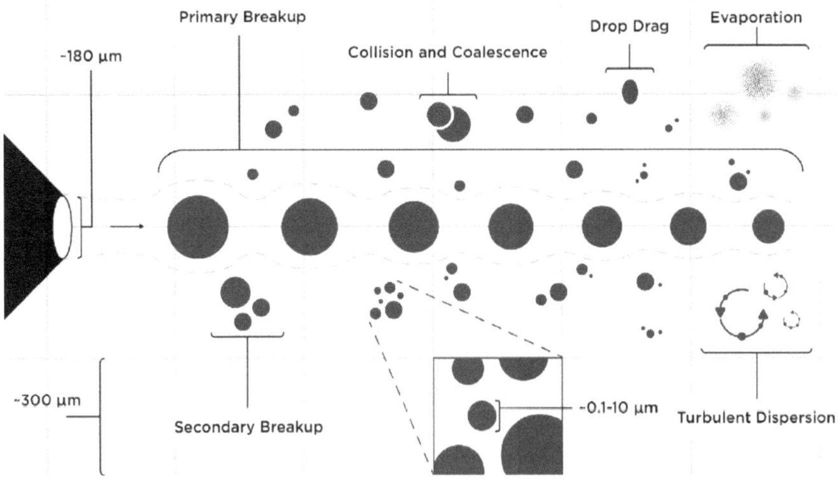

FIGURE 3.29 Physical processes that a liquid fuel parcel undergoes, before evaporating into gaseous species [15].

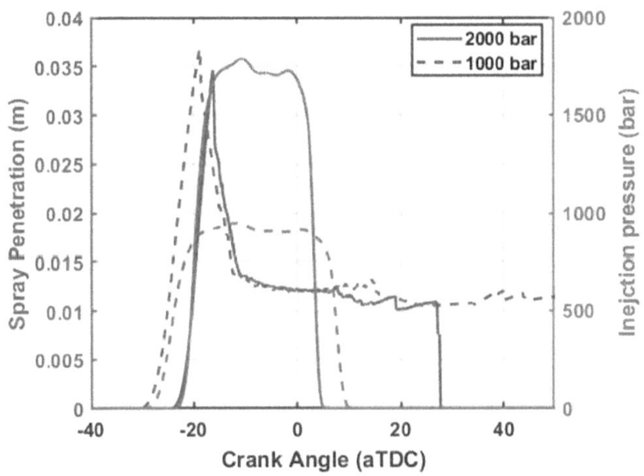

FIGURE 3.30 Spray penetration lengths predicted by the CFD-RANS simulations of the heavy-duty diesel engine at the rated power conditions with a start of injection timing of 29° bTDC using high and low injection pressures.

gradients are formed near the wall boundaries, commonly called the 'boundary layer' where the flow is dominantly laminar. The boundary layer influences the wall heat flux to a great extent. However, the boundary layer in ICEs is very thin and unsteady. Thus, the boundary layer requires a very fine grid near the wall boundaries to capture and resolve the boundary layer in CFD. To reduce the computational cost, the velocity and temperature gradients near the wall are assumed using a sub-grid scale wall function rather than solving the fluid flow in the boundary layer using a very fine grid.

One of the common wall functions developed by Launder and Spalding [23] to assume the near-wall boundary layer is written in equation (3.36). For generalization, the wall function uses dimensionless values for velocity (u^*), and temperature (T^*), which are formulized as a function of normalized distance from the wall (y^+). The distance from the wall (y) is normalized using density (ρ), viscosity (μ), and shear velocity (u_τ) as shown in equation (3.37):

$$u^* = \left\{ \frac{1}{\kappa} \ln\ln\left(9.8y^+\right)\ y^+ > 11.2\ y^+\ y^+ \le 11.2 \right\} \tag{3.36}$$

$$y^+ = \frac{y \rho u_\tau}{\mu} \tag{3.37}$$

Ideally, heat transfer to the wall occurs by conduction from the last fluid molecule near the wall. Since the CFD-RANS simulations assume the boundary layer near the wall instead of solving it, a heat transfer sub-model must be used to calculate the heat flux from the last fluid cell to the wall, assuming equilibrium between them. A commonly used heat transfer sub-model for CFD developed by O'Rourke and Amsden [19] is used in the simulation shown in equation (3.38). The constant F in the equation is used to scale the heat flux based on the distance of the last fluid cell away from the wall, which has a functional form developed based on the near-wall temperature profile as shown in equation (3.39):

$$\dot{q} = K\frac{\partial T}{\partial x} = \frac{\mu_m c_p}{Pr_m} \cdot F \cdot \frac{T_f - T_w}{y} \tag{3.38}$$

$$F = \left\{ 1\ y^+ \left\langle\ 11.05\ \frac{y^+ \dfrac{\mathrm{Pr}_t}{\mathrm{Pr}_m}}{\dfrac{1}{k}\ln\ln\ (y^+) + B + 11.05\left(\dfrac{\mathrm{Pr}_t}{\mathrm{Pr}_m} - 1\right)}\ y^+ \right\rangle 11.05 \right\} \tag{3.39}$$

$\dot{q}$ – heat flux; K – thermal conductivity; μ_m – molecular viscosity; c_p – specific heat; Pr_m –molecular Prandtl number; Pr_t – turbulent Prandtl number; T_f – gas temperature; T_w – wall temperature

3.4.2 Finite Element Analysis

3-D FEA is another commonly used high-fidelity simulation tool for ICE analysis. Thermal and structural analysis of ICE components are often performed using 3-D FE models to predict the durability of ICEs. The underlying principle of the FEA is similar to the CFD, where the control volume is discretized into several small volumes/elements. The energy conservation equations are used to solve for local variables such as temperature, deformation, etc., considering the energy fluxes crossing each boundary. Heat transfer and electrical and structural analyses (sometimes coupled) can be performed using FE models. The energy conservation equation used for uncoupled heat transfer analysis by the Abaqus computer-aided engineering (CAE) software tool is shown in equation (3.40) [24]:

$$\int \rho \dot{U} dV = \int q dS + \int r dV \tag{3.40}$$

where $\dot{U}$ is the rate of change of internal energy of an element, q is the heat flux crossing each face of an element and r is the internal heat generation. The heat flux crossing each face is governed by Fourier's law of conduction as shown in equation (3.41) where k is the thermal conductivity:

$$q = -k \cdot \frac{dT}{dx} \tag{3.41}$$

Proper prescription of boundary conditions is crucial for an FEA to obtain accurate solutions. To conceive the importance of boundary conditions used for FEA, the following sub-section provides a literature review of the boundary conditions used for a 3-D FE thermal analysis of a combustion chamber component. 3-D FE models are commonly used for heat transfer analysis of ICE components. The boundary conditions prescribed vary from simple experimental correlations to 3-D maps from CFD found in literature. Based on the literature review, a methodology developed by synthesizing 3-D CFD and FEA will be elaborated upon in the following sub-section.

3.4.2.1 Literature Review of Boundary Conditions for FE Thermal Analysis

The two major boundary conditions required for an FE thermal model to perform heat transfer analysis in an IC engine are the thermal loads from combustion and the cooling conditions. It should be noted that combustion side boundary conditions can be prescribed as either heat flux or convection as defined by Newton's cooling formula (i.e., heat transfer coefficients and near-wall gas temperatures). Heat transfer by radiation is usually ignored since it is negligible compared to convection or is sometimes included in the heat flux. The most accurate way of defining boundary conditions of an FE thermal analysis would be to use heat flux measurements from ICEs. However, each technique found in the literature has its advantages and disadvantages, which are laid out below.

Heat fluxes measured from an experimental engine using thermocouples were used as the combustion side boundary conditions to the FE piston model to evaluate potential materials for a heavy-duty diesel engine piston [25]. Nevertheless, heat flux measurement in IC engines is time- and cost-intensive. Alternatively, zero-dimensional heat transfer models such as the Woschni [3] or the Hohenberg [4] were used to predict the combustion side boundary conditions for the FE thermal model [26,27,28]. Simple empirical formulae were also used to predict the combustion side boundary conditions [29]. Using zero-dimensional heat transfer models or the empirical formulae for a 3-D thermal analysis is questionable since they involve empirical correlations and lack spatial discretization. The spatially varying transient combustion side boundary conditions could be predicted by simulating combustion using 3-D computational fluid dynamics (CFD).

Fontanesi et al. simulated diesel combustion using CFD and applied the CFD-predicted convective heat transfer boundary conditions for a V6 engine head 3-D FE model to assess fatigue after cycle-averaging (steady-state) [30]. The transients were ignored to save computational expense since the domain is large. Esfahanian et al. performed a thermal analysis of an SI engine piston using a 3-D FE model by prescribing CFD-predicted combustion side boundary conditions [31]. The study provided an understanding of the tradeoff between accuracy and computational time, depending on the processing of the applied combustion side boundary conditions (i.e., fully transient, time-average, or surface-average). It was stated that a spatially varying transient heat transfer analysis would be needed to capture the transient interaction between the heat flux from combustion and the piston temperature with added computational expense. The wall temperature boundary conditions for the initial CFD simulation were usually assumed. Wu et al. applied the surface temperatures predicted by the 3-D FE thermal analysis as boundary conditions for the consequent CFD simulation. They iterated between the CFD and the FE thermal model for better accuracy [32]. The number of iterations depends on the assumed initial temperature boundary conditions for the CFD.

Since temperatures of a solid component require multiple cycles to reach a steady-state, running multiple cycles of 3-D numerical simulations would be cumbersome. Sun et al. used a multi-time-scale approach to optimize the computational time required for 3-D transient heat transfer analysis of an IC engine piston [33]. Moser et al. developed a framework for transient heat transfer predictions on a heavy-duty diesel engine piston by coupling a 3-D CFD and a 3-D FE thermal model offline [34]. The framework offered better control over mesh and time steps individually in the CFD and the FE thermal model since thermal diffusion time scales differ between fluid and solid regions. The authors termed the framework or the methodology as the 'CFD-FEA co-simulation'. Moreover, the CFD-FEA co-simulation was demonstrated to capture any in-cylinder changes caused by the changes in piston surface temperatures. Motwani et al. utilized the same framework to predict the temperature swing of low thermal inertia coatings applied on an SI engine piston [35]. They iterated between the fluid and the solid side predictions for better accuracy. The study used four iterations between the CFD and the FEA for the piston surface temperatures to converge.

Nonetheless, cooling boundary conditions are also required for heat transfer analysis in an IC engine. Due to the cooling conditions' relatively lower significance compared to the combustion side, constant temperature [36] or convection [37] boundary conditions were applied based on sensitivity studies and literature. Some works have used empirical formulae for the cooling conditions [29,31]. For better accuracy, CFD could be used to predict spatially varying cooling conditions needed for a 3-D thermal analysis. Coolant flow through coolant galleries was simulated using CFD, and the convective heat transfer predicted was then used as the cooling boundary conditions for the 3-D thermal analysis of a multi-cylinder diesel engine head to assess fatigue, along with combustion side boundary conditions also predicted by CFD [30]. Some works have simulated oil flow through oil galleries of pistons to predict the cooling phenomenon there [38]. However, the computational expense of performing 3-D CFD for both the combustion and the cooling sides would be onerous. Wright et al. used a Bayesian-based calibration methodology to calibrate the piston cooling heat transfer coefficients (HTCs) based on the CFD-predicted combustion side boundary conditions and measured sub-surface temperatures [39].

It can be understood from the literature review that the combustion side boundary conditions are relatively important for the thermal analysis of a combustion chamber component. The current state-of-the-art is the mapping of convective heat transfer boundary conditions predicted by 3-D CFD. Thus, a synergy of 3-D CFD and 3-D FEA would be valuable for thermal analysis of a combustion chamber component, commonly termed as high-fidelity multi-physics simulation. To illustrate the use of high-fidelity multi-physics simulation in current ICE research, a methodology to co-simulate 3-D CFD and 3-D FEA for heat transfer analysis of a heavy-duty diesel engine piston is detailed below. Following the methodology, its applications are demonstrated by applying it to study the sensitivity of transient heat flux distribution to injection parameters on a heavy-duty diesel engine piston.

3.4.3 3D CFD and FEA Co-simulation

Previously, a reduced-order thermal analysis was carried out by combining the 0-D heat transfer model and a 1-D unsteady heat conduction model. The reduced-order analysis provided some insights about heat transfer in the heavy-duty diesel engine. However, spatial discretization is necessary to capture the strong spatial gradients that exist in diesel engine piston heat transfer. Thus, higher fidelity computational techniques such as three-dimensional computational fluid dynamics (3-D CFD) and 3-D FE thermal models are coupled to capture spatially varying transient heat flux and surface temperature distribution on the heavy-duty diesel engine piston. The framework used for coupling 3-D CFD and FE thermal analysis was developed by Moser et al. [34] termed as 'CFD-FEA co-simulation'. Figure 3.31 shows the process flow chart of the methodology employed and a brief description of the steps are provided as follows:

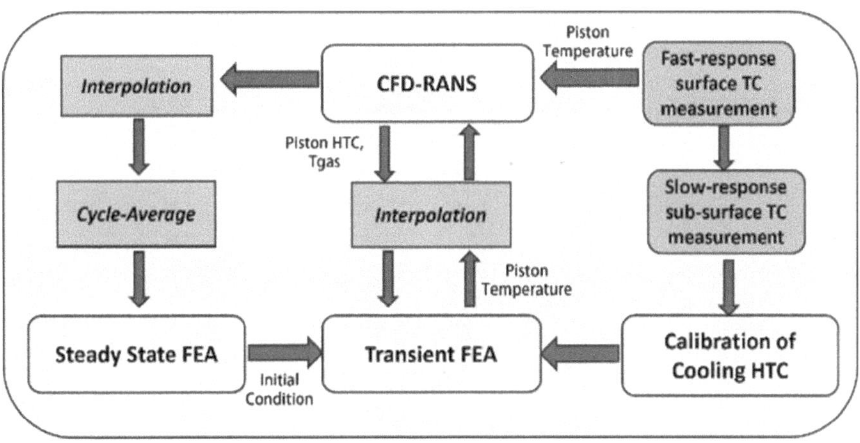

FIGURE 3.31 Process flow chart for the CFD-FEA co-simulation methodology.

Validation: An injection timing (i.e., combustion phasing) sweep at rated power (2500 RPM, 20 bar IMEP in the heavy-duty diesel engine was simulated using CFD. RANS turbulence models were used for the simulation for computational efficiency. The simulations were validated with the measured cylinder pressures and the processed gross heat release rates, as shown in Figure 3.32. A maximum difference of 5 bar between the simulation and the measurement was calculated but is within the cycle-to-cycle differences (maximum of 15 bar) in the cylinder pressure measurements.

Cycle-Averaging and Interpolation: The CFD-predicted crank angle-resolved piston near-wall gas temperatures $\left(T_g\right)$ and convective heat transfer coefficients (h) were interpolated to 4379 evenly spaced nodes on the piston surface as shown in Figure 3.33 to reduce the input file sizes for the FE thermal analysis, and to reduce the processing time. The interpolated T_g and h at each node were then cycle-averaged using equations (3.42) and (3.43) where h_t and T_{g_t} are the instantaneous heat transfer coefficients and gas temperatures respectively. τ is the cycle time, and Δt is the time step:

$$h = \frac{\sum_{t=0}^{\tau} h_t \cdot \Delta t}{\tau} \tag{3.42}$$

$$T_g = \frac{\sum_{t=0}^{\tau} h_t . T_{g_t} \cdot \Delta t}{\sum_{t=0}^{\tau} h_t \cdot \Delta t} \tag{3.43}$$

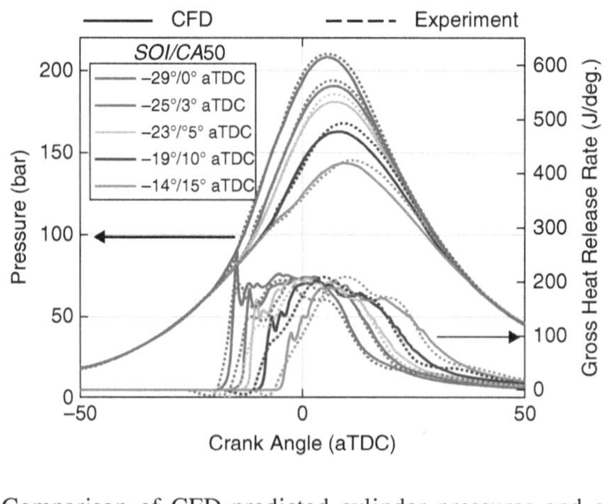

FIGURE 3.32 Comparison of CFD-predicted cylinder pressures and gross heat release rates with measurements for injection timing sweep.

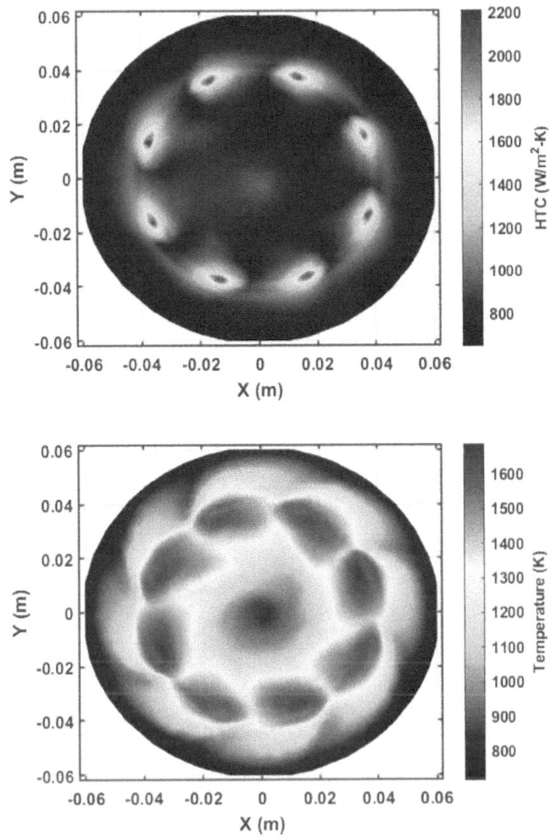

FIGURE 3.33 CFD-predicted cycle-averaged heat transfer coefficients (top) and near-wall gas temperatures (bottom) on the piston surface.

Cooling Condition Calibration: The cycle-averaged T_g and h were applied on the crown surface of the piston FE model. The measured slow-response sub-surface thermocouple temperatures were used to calibrate the backside cooling heat transfer coefficients for the oil gallery and under-crown surfaces shown in Figure 3.34, using a Bayesian model calibration methodology [39].

Steady-state Thermal Analysis: A steady-state thermal analysis was performed using the 3-D piston FE model. The cycle-averaged T_g and h were applied on the piston crown and the calibrated heat transfer coefficients were prescribed as radially constant on the cooling surfaces. Other surface boundary conditions were assumed from the literature [25]. Figure 3.35 shows a cut section of the steady-state piston FE model temperatures for the rated power condition at a start of injection timing of 19° before TDC.

Transient Thermal Analysis: Transient piston thermal analysis was performed using the 3-D FE model to predict the transient, spatially varying surface temperatures. The CFD-predicted crank angle-resolved piston near-wall gas temperatures (T_g) and convective heat transfer coefficients (h) were applied on the crown surface. The backside cooling boundary conditions that were used

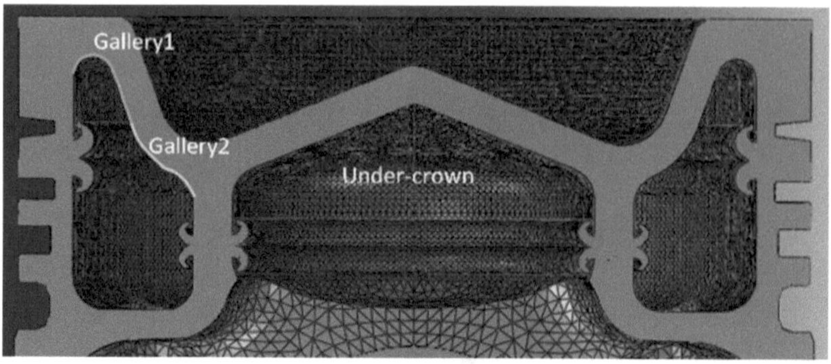

FIGURE 3.34 Piston backside cooling surfaces for which the calibration of HTCs was performed.

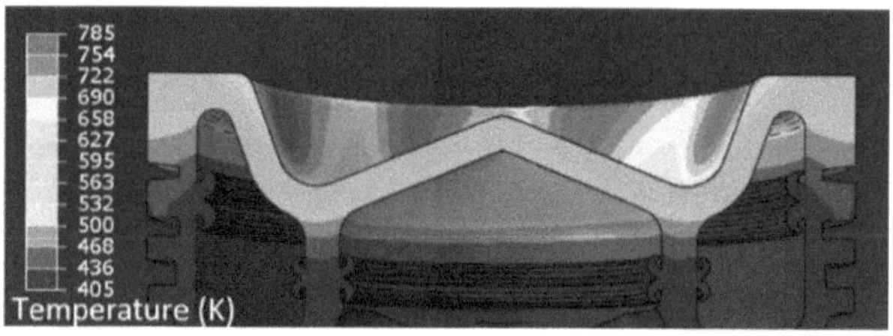

FIGURE 3.35 Steady-state piston temperatures predicted by FE thermal analysis.

were the same as the steady-state analysis. The initial temperature field was prescribed from the steady-state thermal analysis. The consequent cycles were initialized with the last step of the previous cycle. Figure 3.36 shows the transient surface-averaged piston temperatures of three consecutive cycles for the rated

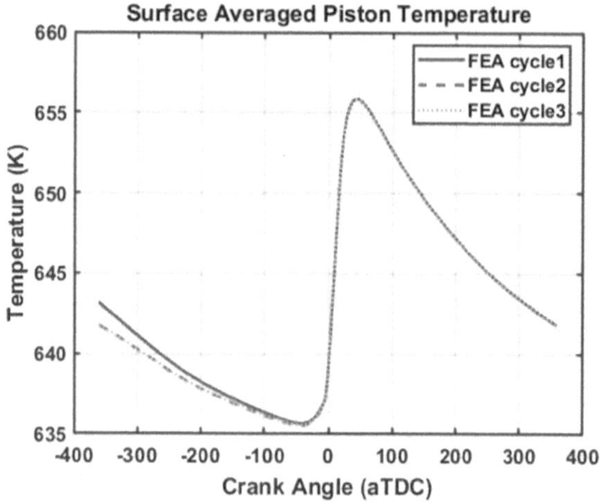

FIGURE 3.36 Transient FEA-predicted crank angle-resolved area-weighted surface-averaged piston temperatures for the three consecutive cycles.

power condition with a start of injection timing of 19° before TDC. The surface nodal temperatures converged in the second cycle, with a maximum error of 0.2 K between the second cycle and third cycle.

3.4.4 Injection Parameter Sensitivity

Experimental work [40,41] using thermocouples embedded in the piston has provided insights into the potential use of injection control parameters, such as injection pressure, timing, etc., to modify piston thermal loads in heavy-duty diesel engines and improve their durability. The durability of surface thermocouples for heat flux measurement and the lack of spatial resolution due to mechanical limitations constrained the scope of heat transfer studies in heavy-duty engines experimentally. The CFD-FEA co-simulation framework described above can be leveraged to study spatially varying transient heat transfer in heavy-duty engines. Transient heat flux and the temperature distribution on a heavy-duty diesel engine piston due to injection timing and pressure were analyzed numerically using the combined high-fidelity simulation methodology and the results are described below. The details of the analysis can be found in a recently published work [42].

3.4.4.1 Injection Timing

Injection timing/combustion phasing was retarded up to 15 degrees from the most advanced injection timing of 29° bTDC. Figure 3.37 shows the combustion duration (CA10–90) and the thermal efficiency as a function of combustion phasing (CA50). The peak thermal efficiency was predicted for CA50 at TDC, and the thermal efficiencies dropped for later combustion phasings due to the elongated combustion duration and increased exhaust enthalpy. A thermal efficiency reduction of approximately 2.5 percentage points was predicted when combustion phasing was retarded 15 degrees. On the other hand, a 75°C reduction in peak surface nodal temperature was predicted for the most retarded combustion phasing, as seen in Figure 3.38. These results show

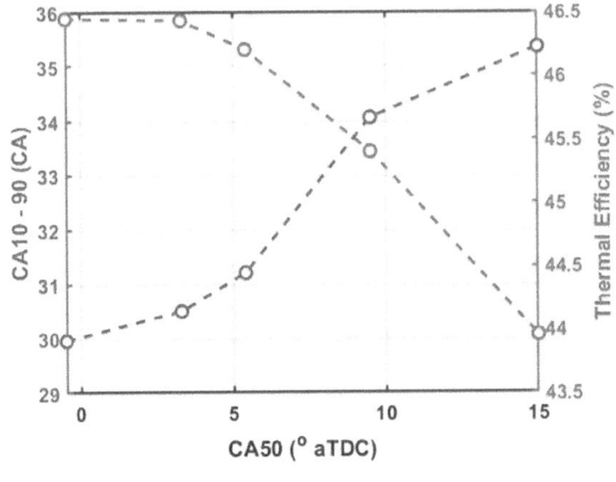

FIGURE 3.37 CFD-predicted thermal efficiency and combustion duration for the injection timing sweep at rated power.

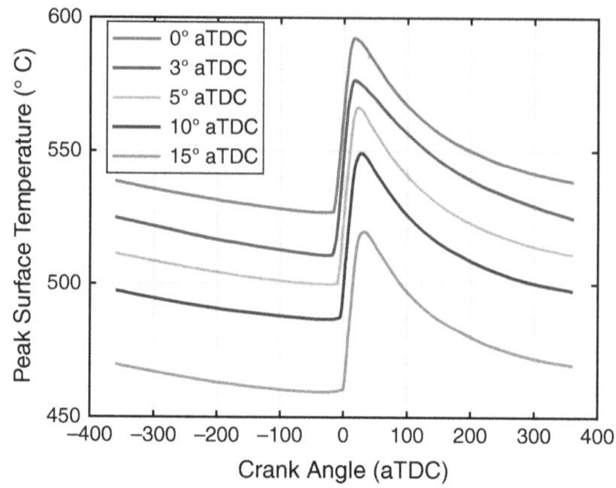

FIGURE 3.38 Maximum temperature predicted by the transient thermal analysis using the piston FE model for the injection timing sweep at rated power.

that there is a tradeoff between thermal efficiency and peak piston temperatures, which could be leveraged to find the optimum combustion phasing for an operating condition that maximizes thermal efficiency without exceeding the thermal limits of the piston.

Figure 3.39 shows the peak surface-averaged heat flux and temperature for the piston center, bowl, and squish regions as a function of CA50. The bowl region absorbed the maximum peak heat flux and showed maximum surface temperature across the combustion phasing sweep since the spray targets the bowl. The center region exhibited the maximum reduction in peak heat flux and surface temperatures when the combustion phasing is retarded because the combustion spreads away from the bowl and the center and into the squish regions for later combustion phasings. The iso-surface plots at CA10, CA50, and CA90 timings in Figure 3.40 for the most advanced and most retarded injection timings show the evolution of combustion with time for the advanced and retarded combustion phasing. Thus, the squish and the center peak heat fluxes are close for later combustion phasings, though the center exhibited significantly higher peak heat fluxes for advanced combustion phasings. The peak surface temperatures are consistently higher for the squish than the center, contrary to the peak heat flux trends. This effect is because the squish region's cooling is limited compared to the bowl region's cooling pathways due to the piston geometry.

3.4.4.2 Injection Pressure

The injection pressure for the combustion phasing sweep was 2000 bar, and the peak thermal efficiency was predicted for the most advanced injection timing. The injection pressure was reduced to 1000 bar for the most advanced injection timing to study the sensitivity of injection pressure. Table 3.3 compares some key injection and performance parameters for the higher and lower injection pressure cases. Due to the reduced injection pressure, the combustion duration was elongated by 9.5°, which contributed to a thermal efficiency reduction of approximately one percentage point. On the other hand, the peak surface temperature is reduced by 17°C due to a significant reduction in heat transfer coefficients for the lower injection pressure case.

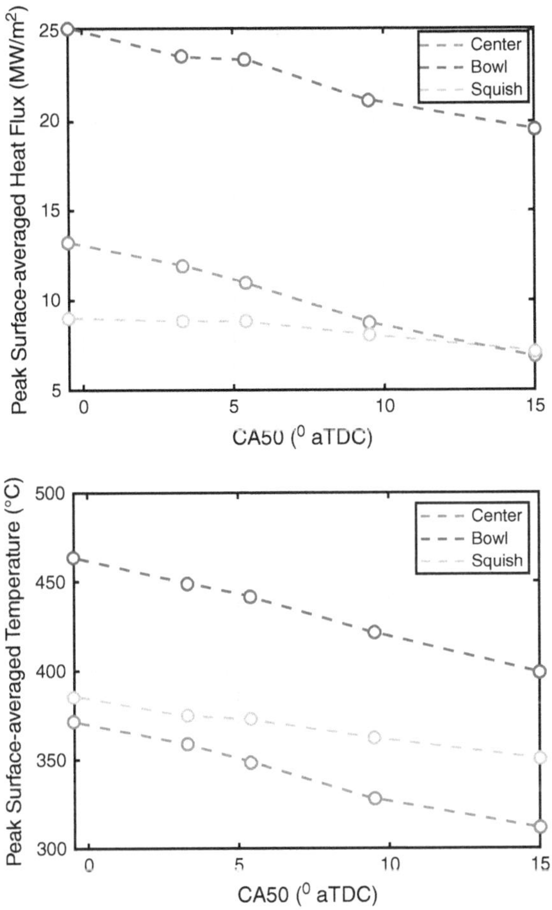

FIGURE 3.39 Peak surface-averaged heat flux (top) and temperatures (bottom) as a function of injection timing predicted for the piston center, bowl, and squish regions.

The results suggest that the injection pressure can be reduced to reduce the piston thermal load with a thermal efficiency penalty but not as significant as the injection timing.

Figure 3.41 shows the surface-averaged transient heat fluxes of the piston center, bowl, and squish regions, along with the gross heat release rates for the lower and higher injection pressure cases. It should be noted that the injection timings are adjusted to obtain the same combustion phasing (CA50). The bowl heat flux rises significantly earlier than the squish and the center since the spray targets the bowl region. However, the peak heat flux is in phase for the bowl and the center since the combustion spreads gradually from the bowl to the center as combustion progresses and then to the squish. A reduction in the peak heat flux of approximately 8 and 2 MW/m^2 in the bowl and the center, respectively, were predicted for the lower injection pressure case. However, the squish heat flux showed no significant difference due to more combustion happening in the squish for an elongated combustion duration. A 17°C, 20°C, and 6°C peak temperature drop were predicted in the center,

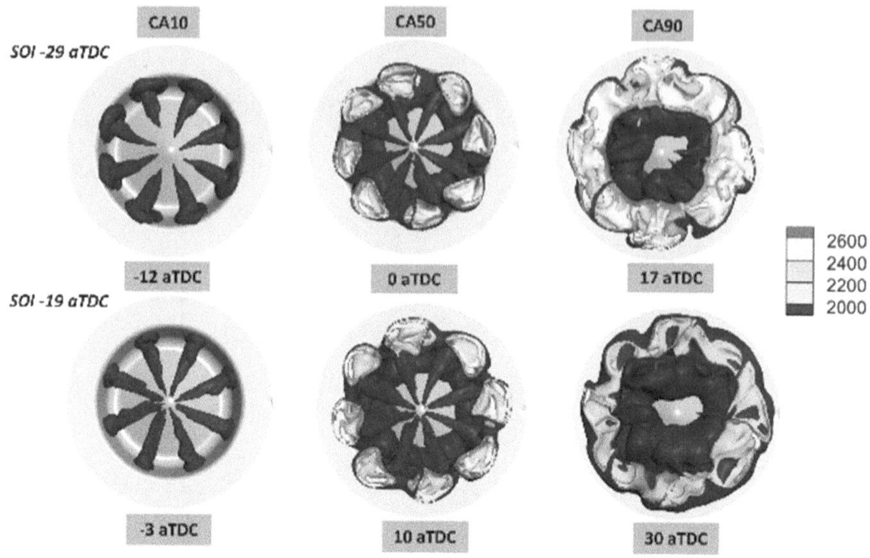

FIGURE 3.40 Temperature iso-surfaces from the CFD simulations for the advanced and retarded injection timings at CA10, CA50, and CA90 timings.

TABLE 3.3
Injection and Combustion Performance Parameters Compared for the High and Low Injection Pressures

Injection pressure (bar)	2000	1000
Injection duration (ms)	1.9	2.7
Peak cylinder pressure (bar)	208	198
CA50 (deg. aTDC)	−0.5	0.8
IMEPg (bar)	19.47	19.2
CA 10–90 (deg.)	30	39.5
Thermal efficiency (%)	46.4	45.5
Combustion efficiency (%)	98.9	98.9
Max piston surface temperature (°C)	592.3	565.2

bowl, and squish regions by reducing the injection pressure by 1000 bar, which can be seen in Figure 3.42. Figure 3.43 shows the cumulative heat energy plotted on a 2-D representation of an 8th of the piston surface, which shows a narrower peak heat transfer area and a more spread-out heat transfer area for the lower injection pressure.

3.5 COMPUTATIONAL EXPENSE

The tradeoff between the computational expense and the fidelity/accuracy is the primary selection criteria for mathematical models. A one-to-one comparison between the reduced-order and high-fidelity models will not make sense since the high-fidelity

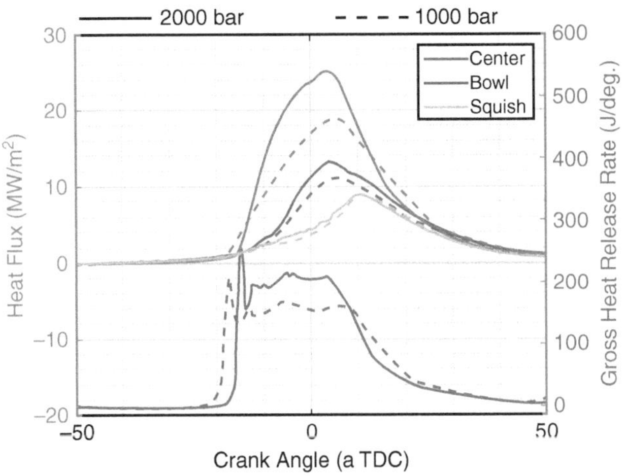

FIGURE 3.41 Transient surface-averaged heat fluxes predicted for the piston center, bowl, and the squish regions along with the gross heat release rates for the two injection pressures.

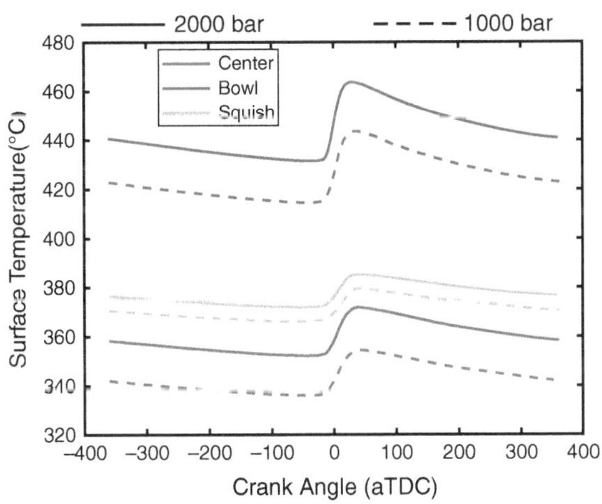

FIGURE 3.42 Transient surface-averaged temperatures predicted for the piston center, bowl, and the squish regions.

simulations require high-performance computers with multiple processors, whereas reduced-order models can be solved with a single core in a local machine. The zero-dimensional and one-dimensional models used for the reduced-order piston thermal analysis discussed in the first section were solved in seconds in a single core of a local machine. Therefore, a discussion related to computational expense for the reduced-order models is ignored.

Even the GT-Suite simulations of the single-cylinder engine models for steady-state operating conditions can be solved in seconds in a local machine. Discussion about computational expense can be considered for the multi-cylinder engine cold-start

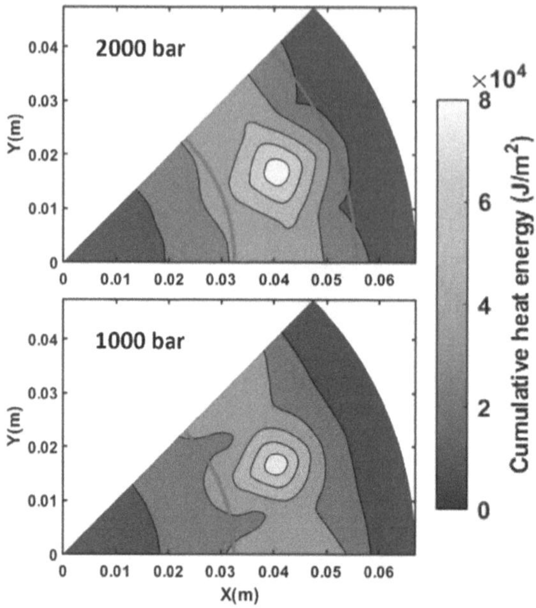

FIGURE 3.43 Cumulative heat energy on 8th of the piston surface for the two injection pressures.

simulations illustrated in the second section since transient thermal solvers are used along with flow solvers. The total volume of the multi-cylinder engine used for the analysis was 8.44L discretized into 240 sub-volumes. The total heat transfer surface area subjected to the transient thermal solver was 3.8 m^2. The total length of the simulated cold-start was 45 seconds (600 engine cycles) using a time step of approximately 0.1 crank angle degrees. Each case of the cold-start simulation used approximately 1 hour in a local machine. The computational time can be greatly reduced using the distributed computing capabilities of GT-Suite on multiple processors in a local machine or a high-performance computer.

A real discussion about computational expense would be necessary for high-fidelity simulations. The number of cells used in the CFD-RANS simulations of the single-cylinder heavy-duty diesel engine ranged from 150,000 up to 3 million throughout the cycle. Twelve hours of computational time were required to simulate one cycle of CFD-RANS using 112 cores. Since the first cycle of CFD-RANS heavily depends on the initial conditions, the convective boundary conditions were extracted from the second cycle. Thus, the total computational time of CFD-RANS simulation was 24 hours. The piston FE model contains approximately 2 million elements and requires 2.5 hours of computational time on 112 cores for one cycle. As previously stated, the predicted surface nodal temperatures converged within a tolerance of 0.2 K in 2 cycles. Thus, the total computational time for a transient FEA simulation was 5 hours. In total, 29 hours of computation time was required to solve one iteration of CFD and FEA on 112 cores.

3.6 CALIBRATION OF REDUCED-ORDER MODEL WITH HIGH-FIDELITY SIMULATION DATA

A mathematical model for any analysis should be selected based on a tradeoff between the accuracy and the computational expense. The higher-fidelity models, such as the 3-D CFD and 3-D FE models, are known for their accurate solutions involving spatial discretization. In contrast, the reduced-order models are known for their computational efficiency with a compromise in accuracy. Nonetheless, the accuracy of the reduced-order models can be improved through calibration using experimental data, but it could be cost-intensive to have detailed measurements from the experiment. An alternate technique is using high-fidelity simulation data to calibrate the reduced-order models.

To demonstrate this approach, the 0-D heat transfer model in GT-Suite is calibrated using CFD-RANS data. In the GT-Suite simulations, wall heat transfer in diesel engines exhibits strong spatial gradients that cannot be captured using 0-D heat transfer correlations. However, the spatial discretization of the CFD simulations can be leveraged to induce spatial variation in GT-Suite heat transfer predictions. Convective heat transfer predictions in the combustion chamber require the gas temperatures and the heat transfer coefficients. GT-Suite uses a two-zone method for gas temperatures where the burned and unburned zone gas temperatures are calculated separately from pressures using the ideal gas law. The heat transfer coefficients (h) are calculated using 0-D heat transfer correlations, and the heat flux from the burned and unburned zones are combined using equation (3.44) where α denotes the mass fraction burned. Instead of using 0-D heat transfer correlations, the CFD-precited heat transfer coefficients were prescribed. The combustion chamber-average heat transfer coefficients predicted by the CFD-RANS simulation of the diesel engine and h using density (ρ) and mean piston speed (v_p) for the rated power condition at a start of injection timing of 29° bTDC are plotted in Figure 3.44. It should be noted that

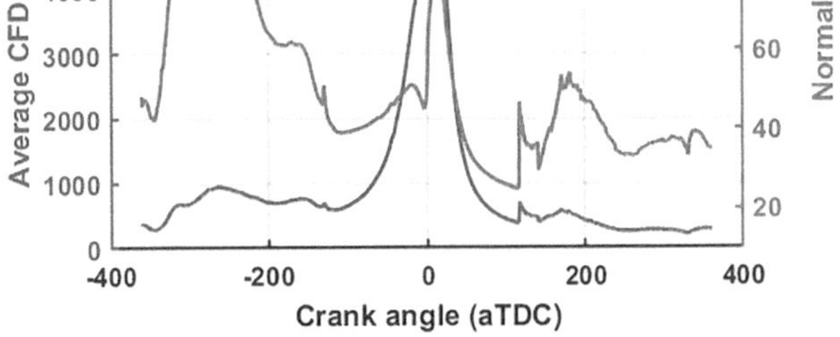

FIGURE 3.44 Combustion chamber-average heat transfer coefficients predicted by CFD and the normalized heat transfer coefficients prescribed in GT-Suite.

the heat transfer coefficients predicted by CFD cannot be directly applied in a 0-D simulation since the CFD simulation uses near-wall gas temperatures to calculate heat flux, whereas a 0-D heat transfer model requires bulk gas temperature.

Figure 3.45 shows the near-wall gas temperature of each combustion chamber surface of the heavy-duty diesel engine predicted by CFD-RANS compared to the burned and unburned bulk gas temperatures predicted by GT-Suite. Thus, the heat transfer coefficients of each surface need to be scaled considering differences between near-wall gas temperatures of each surface and the bulk gas temperature. The individual surface convection multipliers in GT-Suite were used to scale and induce the spatial variations predicted by the CFD. The individual combustion chamber surface

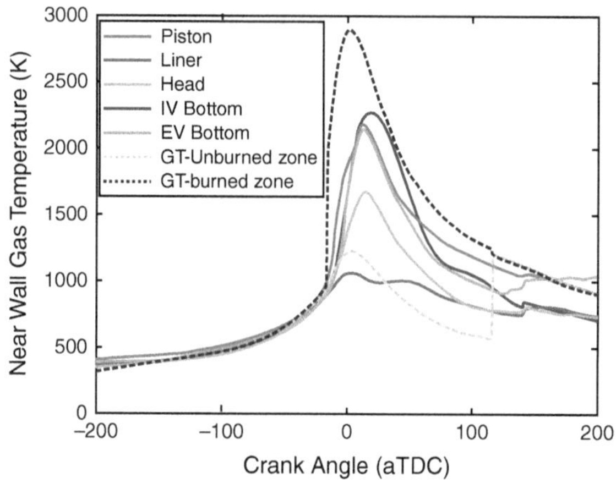

FIGURE 3.45 Near-wall gas temperatures predicted by CFD, and bulk gas temperatures (unburned and burned) predicted by GT-Suite.

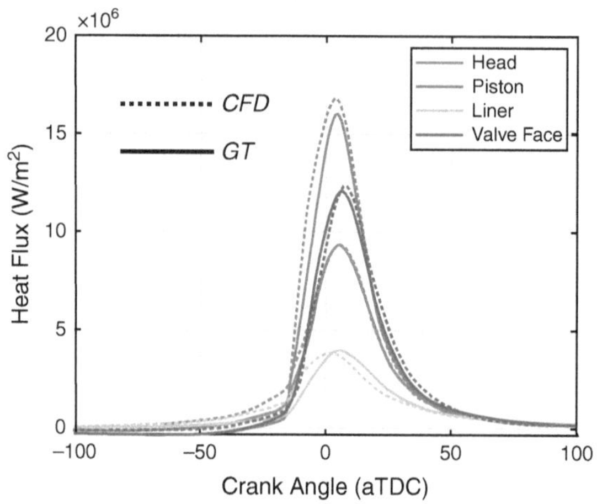

FIGURE 3.46 Combustion chamber surface heat fluxes predicted by CFD and the calibrated GT-Suite model.

heat fluxes predicted by the CFD-RANS and the calibrated GT-Suite predictions are plotted in Figure 3.46.

$$Q = \alpha Ah(T_b - T_w) + (1-\alpha)Ah(T_u - T_w) \tag{3.44}$$

Q – heat flux; A – surface area; h – heat transfer coefficient; T_b – burned gas temperature; T_u – unburned gas temperature; T_w – wall temperature

3.7 NUMERICAL TOOLS FOR ASSESSING FUTURE ICES

The future of ICEs depends on emissions reductions and efficiency improvements that are feasible. This section provides an overview of the application of numerical tools to aid the research community in achieving its goals. The ICEs used for modern automotive applications are dominantly two types: compression ignition (CI) and spark-ignition (SI) four-stroke engines. The CI engines exhibit high thermal efficiency due to their high compression ratios with lean and unthrottled operation along with relatively low combustion temperatures. However, CI engines produce high soot emissions from locally rich fuel-air mixtures (high equivalence ratios). NOx emissions that are produced from a CI engine are difficult to treat due to the lean operation that contributed to the high efficiency. SI engines produce low soot emissions due to the premixed mixtures used but suffer from low efficiencies due to their relatively lower compression ratios with stoichiometric and throttled operation. Additionally, SI engines incur high thermal losses and NOx emissions due to high combustion temperatures.

A tradeoff between soot and NOx exists in CI engines, where varying one control parameter often results in one harmful pollutant emission being traded for the other. Analogously, a proverbial soot and NOx tradeoff exists between CI and SI engines due to the nature of the CI and SI combustion process, which can be better understood from the equivalence ratio-temperature plot in Figure 3.47. A favorable environment for soot emissions is a high equivalence ratio and modest temperature,

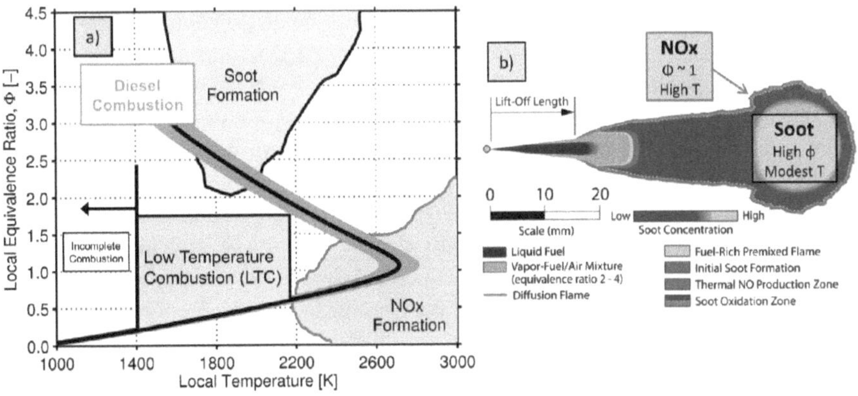

FIGURE 3.47 (a) Local equivalence ratio-temperature maps showing NOx and soot forming regions, (b) Typical diesel combustion flame showing soot and NOx formation regions [43].

whereas the NOx emissions are favored in near-stoichiometric equivalence ratio, high-temperature regions. Finally, a global efficiency-emissions tradeoff exists between SI and CI, where SI results in generally favorable tailpipe emissions but with lower efficiencies, and CI results in generally favorable efficiencies but with worst emissions characteristics.

Novel combustion processes are being developed to break these tradeoffs by navigating the soot and NOx production regions on the equivalence ratio-temperature plot in Figure 3.47. These strategies are commonly known as low-temperature combustion. The low-temperature combustion strategies operate at lower peak temperatures than SI or CI, and at generally lower equivalence ratio regions to avoid producing soot and NOx. Moreover, low combustion temperatures reduce heat losses to coolant through combustion chamber walls by maintaining lower peak temperatures which helps improve efficiency. Finally, the higher compression ratios that are possible with low-temperature combustion and the lean and unthrottled operation cause low-temperature combustion to have diesel-like or higher-than-diesel efficiencies.

Homogeneous charge compression ignition (HCCI) is one of the earliest low-temperature combustion strategies that broke the efficiency-emissions tradeoff and the soot-NOx tradeoff by operating in low equivalence ratio, low-temperature regions. This is achieved by auto-igniting a homogeneous lean mixture through compression heating alone, which is a blend between the favorable aspects of the SI and CI combustion processes (i.e., premixed operation paired with lean and unthrottled operation with high compression ratios). The challenge with HCCI is the narrow operating range due to rapid heat release rates. Moreover, the ignition in HCCI is driven purely by the chemical kinetics and the compression heating, and therefore, the ignition is difficult to control. Thermal stratification [44], reactivity stratification [45], equivalence ratio stratification [46], etc. are proposed as potential ways to stagger the heat release process in low-temperature combustion and broaden the operating range. GT-Suite has introduced some combustion models to analyze advanced combustion modes, such as HCCI, where a chemical kinetics solver is used. However, 0-D combustion models will not be helpful for stratified combustion studies where spatial resolution is required. There is a 'stochastic reactor model' in GT-Suite which is a multi-zone combustion model that could capture some level of stratification, but the accuracy is questionable. 3-D CFD is considered the optimal tool for stratified combustion studies.

To illustrate the potential use of 3-D CFD for stratification studies, an HCCI-style combustion process is simulated in a single-cylinder medium-duty engine with a compression ratio 15.5 using 3-D CFD. Wet ethanol 80 (i.e., 80% ethanol, 20% water) is injected during the intake stroke at 270° and 240° before combustion TDC [47]. Figure 3.48 shows the mass joint probability density function of equivalence ratio and temperature before ignition (10° bTDC), which is helpful to visualize the stratification differences due to the difference in injection timing. The later injection timing stratified the charge in the cylinder more than the earlier injection timing due to there being less time available for mixing. Figure 3.49 shows a staggered gross heat release rate for the later injection compared to the earlier injection due to the increased stratification. Wet ethanol 80 was preferred due to its high latent heat of vaporization that could promote a controlled level of temperature stratification using varying injection timing.

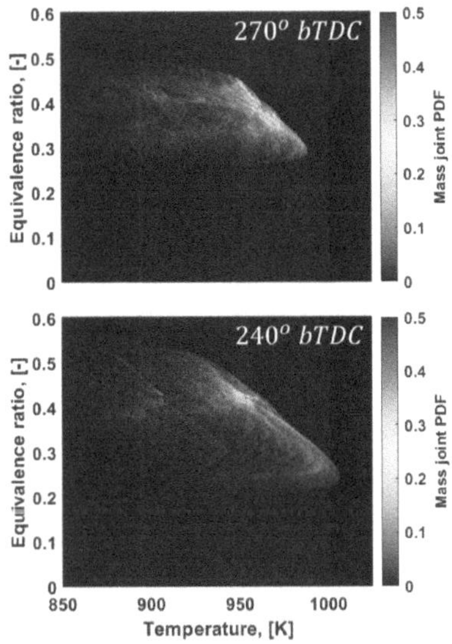

FIGURE 3.48 Mass joint probability density function of temperature and equivalence ratio before ignition (10° bTDC) predicted by 3-D CFD for the two intake stroke injections.

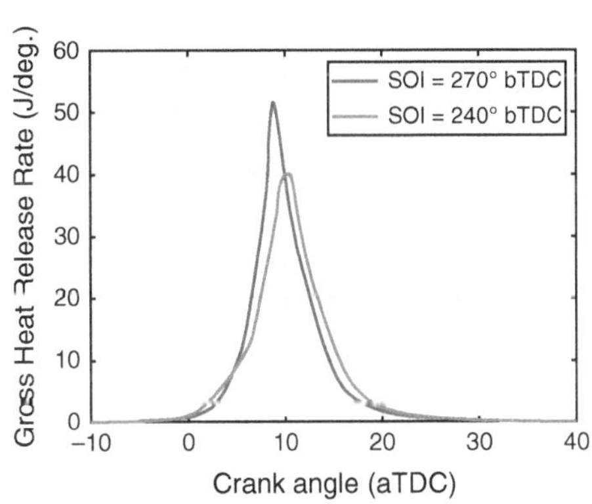

FIGURE 3.49 Gross heat release rates predicted by CFD for the two intake stroke injections.

Moreover, other means of stratification, such as residual gas fraction, fuels, etc. can be studied using 3-D CFD. Renewable fuels such as alcohol fuels [48], hydrogen [49], natural gas [50], etc. have gained traction recently, and 3-D CFD is a widely used tool to aid research in these areas. Two-stroke engines were previously more popular than they are currently. The advantage of a two-stroke engine is its higher power density compared to four-stroke engines. But two-stroke engines lost popularity in certain sectors over the years due to their increased emissions. The main issue with two-stroke engines is their breathing (i.e., scavenging) inefficiencies. An optimized flow path for intake and exhaust could improve scavenging and, consequently, the emissions of a two-stroke engine. 3-D CFD can be used to optimize the flow path designs of two-stroke engines [51], reducing the burden on experimental studies.

Machine learning/artificial intelligence has been gaining popularity in the research community. Researchers are attempting to adopt data-driven machine learning techniques for accelerating ICE research. High-fidelity ICE simulations are data and time intensive. Machine learning techniques like neural networks are being explored to aid high-fidelity simulations and reduce computational expense [52]. However, since most machine learning techniques are data-driven, they require a large amount of input data for them to 'learn' the correlations and 3-D CFD is already computationally expensive. Additionally, researchers are skeptical about the data-driven model's disconnect from physics. This concern can be addressed by the physics-guided machine learning approaches [53], which are currently explored for aiding high-fidelity ICE simulations. With that said, the machine learning approaches can help accelerate ICE research in the future.

3.8 CONCLUSION

The advancement of computing technologies has enabled the ICE research community to use mathematical models and reduce the reliance on experimental studies. Mathematical models used for ICE research vary from reduced-order to multi-dimensional higher fidelity models. The choice of model for analysis should be made by considering the objective of the study and each model's ability to capture the physics, as well as the tradeoff between accuracy and computational expense. Various fidelities of mathematical models used in ICE research and their applications to ICEs were laid out by presenting case studies to understand the differences and aid in model selection. The reduced-order models are easily executable in local machines and can provide preliminary insights. However, they lack spatial resolution, which could be crucial for specific studies. Moreover, reduced-order models rely primarily on experimental data.

Multi-physics simulation tools such as GT-Suite reduce the need for experimental data by synthesizing various reduced-order models. The computational expense of multi-physics simulations is affordable to run on local machines, but distributed computing on multiple cores could accelerate the simulations. Coarse spatial resolution can be used in GT-Suite to capture spatial variations and stratification. Experimental data would be required for calibration of the sub-models in GT-Suite. Overall, multi-physics ICE simulations are valuable for system-level studies where high-fidelity models are expensive but suffer from the same drawbacks as the reduced-order models.

Multi-dimensional numerical tools, such as 3-D CFD and FEA, could accurately reproduce experimental results to the minutest details. High-fidelity models are now affordable for ICE simulations due to high-performance computing. Nonetheless, certain numerical techniques considering some assumptions (like turbulence sub-models) still need to be applied in high-fidelity models to run full ICE simulations. Thus, the model selection should depend on the tradeoff between the accuracy required for analysis and the computational power available. One way to improve the tradeoff is to calibrate the reduced-order models with high-fidelity simulations. Numerical tools are becoming more relevant due to stringent emission norms and accelerated vehicle development timelines. Additionally, numerical tools are impactful for research on novel combustion modes, alternate fuels, and alternate engine architectures that are the key to developing better ICEs.

REFERENCES

1. Nusselt, W. (1923). Der Wameubergang in der Verbrennungs-krafimaschine. *Zeitschrift des Vereins deutscher Ingenieure*, *67*, p.708.
2. Elser, K. (1954). *Der instationäre Wärmeübergang in Dieselmotoren: theoretische und experimentelle Untersuchungen*. Leemann, Zurich.
3. Woschni, G. (1968). A universally applicable equation for the instantaneous heat transfer coefficient in the internal combustion engine. *SAE Transactions*, pp. 3065–3083. https://doi.org/10.4271/670931
4. Hohenberg, G. F. (1979). Advanced approaches for heat transfer calculations. *SAE Transactions*, pp. 2788–2806. https://doi.org/10.4271/790825
5. Dao, K., O. A. Uyehara, and P. S. Myers. (1973). Heat transfer rates at gas-wall interfaces in motored piston engine. *SAE Transactions*, pp. 2237–2258. https://doi.org/10.4271/730632
6. Pillai, A. L., R. Kai, T. Murata, T. Ikedo, R. Masuda, and R. Kurose. (2022). Numerical analysis of heat transfer characteristics of spray flames impinging on a wall under CI engine-like conditions. *Combustion and Flame*, *239*, p. 111615. https://doi.org/10.1016/j.combustflame.2021.111615
7. Gohn, J., E. Gingrich, M. Tess, V. Korivi, Z. Yan, B. Gainey, Z. Filipi, and B. Lawler. (2023). *Thermodynamic Modeling of Military Relevant Diesel Engines with 1-D Finite Element Piston Temperature Estimation. No. 2023-01-0103*. SAE Technical Paper. https://doi.org/10.4271/2023-01-0103
8. Gingrich, E. M. (2020). *High-output Diesel Engine Heat Transfer*. The University of Wisconsin-Madison. https://apps.dtic.mil/sti/citations/AD1109699
9. GT-Suite Manual (2024). *Engine Performance and Flow Theory Manual*. Gamma Technologies, Westmont, IL, USA.
10. Colburn, A. P. (1964). A method of correlating forced convection heat-transfer data and a comparison with fluid friction. *International Journal of Heat and Mass Transfer*, *7*(12), pp. 1359–1384. https://doi.org/10.1016/0017-9310(64)90125-5
11. Gandolfo, J., B. Gainey, Z. Yan, C. Jiang, R. Kumar, E. H. Jordan, Z. Filipi, and B. Lawler. (2023). Low thermal inertia thermal barrier coatings for spark ignition engines: an experimental study. *International Journal of Engine Research*, p. 14680874221149458. https://doi.org/10.1177/14680874221149458
12. Roberts, A., R. Brooks, and P. Shipway. (2014). Internal combustion engine cold-start efficiency: a review of the problem, causes and potential solutions. *Energy Conversion and Management*, *82*, pp. 327–350. https://doi.org/10.1016/j.enconman.2014.03.002
13. R. J. Farrauto and R. M. Heck. (1999). Catalytic converters: state of the art and perspectives, *Catalysis Today*, *51*(3–4), pp. 351–360, ISSN 0920-5861, https://doi.org/10.1016/S0920-5861(99)00024-3.
14. Ravikumar, A., A. Bhatt, B. Gainey, and B. Lawler. (2023). *GT-Suite Modeling of Thermal Barrier Coatings in a Multi-Cylinder Turbocharged DISI Engine for Catalyst Light-Off Delay Improvement. No. 2023-01-1602*. SAE Technical Paper. https://doi.org/10.4271/2023-01-1602
15. Richards, K. J., P. K. Senecal, and E. Pomraning (2024). CONVERGE 3.1 Manual, Convergent Science, Madison, WI.
16. Argyropoulos, C. D., and N. C. Markatos. (2015). Recent advances on the numerical modelling of turbulent flows. *Applied Mathematical Modelling*, *39*(2), pp. 693–732. https://doi.org/10.1016/j.apm.2014.07.001
17. Pomraning, E., K. Richards, and P. K. Senecal. (2014). *Modeling Turbulent Combustion using a RANS Model, Detailed Chemistry, and Adaptive Mesh Refinement. No. 2014-01-1116*. SAE Technical Paper. https://doi.org/10.4271/2014-01-1116

18. Reitz, R. D., and Rm. Diwakar. (1987). Structure of high-pressure fuel sprays. SAE Transactions, pp. 492–509. https://doi.org/10.4271/870598
19. Amsden, A. A., P. J. O'Rourke, and T. D. Butler. (1989). *KIVA-II: A Computer Program for Chemically Reactive Flows with Sprays. No. LA-11560-MS.* Los Alamos National Lab (LANL), Los Alamos, NM (United States). https://doi.org/10.2172/6228444
20. Schmidt, D. P., and C. J. Rutland. (2000). A new droplet collision algorithm. *Journal of Computational Physics, 164*(1), pp. 62–80. https://doi.org/10.1006/jcph.2000.6568
21. Liu, A. B., Mather, D., and Reitz, R. D. (1993). Modeling the effects of drop drag and breakup on fuel sprays. *SAE Transactions, 102*, pp. 83–95. https://www.jstor.org/stable/44611358. Accessed 6 Nov. 2023.
22. O'Rourke, P. J., and A. A. Amsden. (2000). A spray/wall interaction submodel for the KIVA-3 wall film model. *SAE Transactions*, 109, pp. 281–98. https://www.jstor.org/stable/44634219. Accessed 6 Nov. 2023.
23. Launder, B. E., and D. B. Spalding. (1983). The numerical computation of turbulent flows. In *Numerical Prediction of Flow, Heat Transfer, Turbulence and Combustion* (pp. 96–116). Pergamon. https://doi.org/10.1016/B978-0-08-030937-8.50016-7
24. Smith, M. (2009). *ABAQUS/Standard User's Manual, Version 6.9.* Dassault SystèmesSimulia Corp.
25. Gingrich, E., D.T. Pierce, G. Byrd, K. Sebeck, V. Korivi, G. Muralidharan, H. Wang, J. Torres, A. Trofimov, J.A. Haynes, and M. Tess. (2022). *Evaluation of High-Temperature Martensitic Steels for Heavy-Duty Diesel Piston Applications. SAE Technical Paper Series* 2022.01. https://doi.org/10.4271/2022-01-0599
26. Baldissera, P., and C. Delprete. (2019). Finite element thermo-structural methodology for investigating diesel engine pistons with thermal barrier coating. *SAE International Journal of Engines, 12*(1), pp. 69–78. https://doi.org/10.4271/03-12-01-0006
27. Mitianiec, W. (2018). Study of influence of boundary conditions on deformation and stresses in a cooled piston of a diesel engine–part B-calculations. IOP Conference Series: Materials Science and Engineering, *421*(4). https://doi.org/10.1088/1757-899X/421/4/042057
28. Gehlot, R., and B. Tripathi. (2016). Thermal analysis of holes created on ceramic coating for diesel engine piston. *Case Studies in Thermal Engineering, 8*, pp. 291–299. https://doi.org/10.1016/j.csite.2016.08.008
29. Lu, Y., X. Zhang, P. Xiang, and D. Dong. (2017). Analysis of thermal temperature fields and thermal stress under steady temperature field of diesel engine piston. *Applied Thermal Engineering*, 113, pp. 796–812. https://doi.org/10.1016/j.applthermaleng.2016.11.070
30. Fontanesi, S., and M. Giacopini. (2013). Multiphase CFD–CHT optimization of the cooling jacket and FEM analysis of the engine head of a V6 diesel engine. *Applied Thermal Engineering, 52*(2), pp. 293–303.
31. Esfahanian, V., A. Javaheri, and M. Ghaffarpour. (2006). Thermal analysis of an SI engine piston using different combustion boundary condition treatments. *Applied Thermal Engineering*, 26(2–3), pp. 277–287.
32. Wu, M., Y. Pei, J. Qin, X. Li, J. Zhou, Z.S. Zhan, Q. Y. Guo, B. Liu, and T. G. Hu. (2017). *Study on Methods of Coupling Numerical Simulation of Conjugate Heat Transfer and In-cylinder Combustion Process in GDI Engine.* No. 2017-01-0576. SAE Technical Paper. https://doi.org/10.4271/2017-01-0576
33. Sun, C., B. Deng, J. Yang, R. Feng, and C. Chen. (2023). A multi-time scales semi-decoupled CHT (Coupled Heat Transfer) model and its application on piston transient heat transfer simulation. *Applied Thermal Engineering, 229*, pp. 120548. https://doi.org/10.1016/j.applthermaleng.2023.120548
34. Moser, S., K. D. Edwards, T. Schoeffler, and Z. Filipi. (2021). CFD/FEA co-simulation framework for analysis of the thermal barrier coating design and its impact on the HD diesel engine performance. *Energies, 14*(8), p. 2044. https://doi.org/10.3390/en14082044

35. Motwani, R., J. Gandolfo, B. Gainey, A. Levi, S. Moser, Z. Filipi, and B. Lawler. (2022). *Assessing the Impact of a Novel TBC Material on Heat Transfer in a Spark Ignition Engine through 3D CFD-FEA Co-simulation Routine. No. 2022-01-0402.* SAE Technical Paper. https://doi.org/10.4271/2022-01-0402
36. Kundu, P., R. Scarcelli, S. Som, Y. Wang, J. Kiedaisch, and M. Rajkumar. (2016). *Modeling Heat Loss through Pistons and Effect of Thermal Boundary Coatings in Diesel Engine Simulations using a Conjugate Heat Transfer Model.* SAE Technical Paper 2016-01-2235. https://doi.org/10.4271/2016-01-2235
37. Mizuno, H., K. Ashida, A. Teraji, K. Ushijima, and S. Takemura. (2009). Transient analysis of the piston temperature with consideration of in-cylinder phenomena using engine measurement and heat transfer simulation coupled with three-dimensional combustion simulation. *SAE International Journal of Engines*, *2*(1), pp. 83–90. https://doi.org/10.4271/2009-01-0187
38. Jin-feng, P. A. N., R. Nigro, and E. Matsuo. (2005). *3-D Modeling of Heat Transfer in Diesel Engine Piston Cooling Galleries [C]. SAE Paper*: 01-1644. https://doi.org/10.4271/2005-01-1644
39. Wright, S., A. Ravikumar, L. Redmond, B. Lawler, M. Castanier, E. Gingrich, and M. Tess. (2023). *Data Reduction Methods to Improve Computation Time for Calibration of Piston Thermal Models. No. 2023-01-0112.* SAE Technical Paper. https://doi.org/10.4271/2023-01-0112
40. Gingrich, E., M. Tess, V. Korivi, and J. Ghandhi. (2022). High-output diesel engine heat transfer: part 1 – comparison between piston heat flux and global energy balance. *International Journal of Engine Research*, *23*(8), pp.1417–1431. https://doi.org/10.1177/14680874211017032
41. Tess, M. J., V. Korivi, E. Gingrich, and P. Schihl. (2023). Influence of spray and combustion processes on piston temperatures and engine heat transfer in a high-output diesel engine. *International Journal of Engine Research*, *24*(5), pp. 2260–2278. https://doi.org/10.1177/14680874221117889
42. Ravikumar, A., S. Wright, L. Redmond, E. Gingrich, V. Korivi, M. Tess, J. Piehl, and B. Lawler. (2024). *Numerical Evaluation of Injection Parameters on Transient Heat Flux and Temperature Distribution of a Heavy-duty Diesel Engine Piston. No. 2024-01-2688.* SAE Technical Paper.
43. Dempsey, A. B., S. J. Curran, and R. M. Wagner. (2016). A perspective on the range of gasoline compression ignition combustion strategies for high engine efficiency and low NOx and soot emissions: effects of in-cylinder fuel stratification. *International Journal of Engine Research*, *17*(8), pp. 897–917. https://doi.org/10.1177/1468087415621805
44. Lawler, B., S. Mamalis, S. Joshi, J. Lacey, O. Guralp, P. Najt, and Z. Filipi. (2017). Understanding the effect of operating conditions on thermal stratification and heat release in a homogeneous charge compression ignition engine. Applied *T*hermal *E*ngineering, *112*, pp. 392–402. https://doi.org/10.1016/j.applthermaleng.2016.10.056
45. Kokjohn, S., R. Hanson, D. Splitter, J. Kaddatz, and R. Reitz. (2011). Fuel reactivity controlled compression ignition (RCCI) combustion in light-and heavy-duty engines. *SAE International Journal of Engines*, *4*(1), pp. 360–374. https://doi.org/10.4271/2011-01-0357
46. Tsolakis, A., and A. Megaritis. (2005). Partially premixed charge compression ignition engine with on-board H2 production by exhaust gas fuel reforming of diesel and biodiesel. *International Journal of Hydrogen Energy*, *30*(7), pp. 731–745. https://doi.org/10.1016/j.ijhydene.2004.06.013
47. O'Donnell, P. C., M. R. Boldaji, B. Gainey, and B. Lawler. (2020). *Varying Intake Stroke Injection Timing of Wet Ethanol in LTC. No. 2020–01–0237.* SAE Technical Paper. https://doi.org/10.4271/2020-01-0237

48. Gainey, B., Z. Yan, and B. Lawler. (2021). Autoignition characterization of methanol, ethanol, propanol, and butanol over a wide range of operating conditions in LTC/HCCI. *Fuel*, *287*, p. 119495. https://doi.org/10.1016/j.fuel.2020.119495
49. Verhelst, S., and T. Wallner. (2009). Hydrogen-fueled internal combustion engines. *Progress in Energy and Combustion Science*, *35*(6), pp. 490–527. https://doi.org/10.1016/j.pecs.2009.08.001
50. Ran, Z., D. Hariharan, B. Lawler, and S. Mamalis. (2020). Exploring the potential of ethanol, CNG, and syngas as fuels for lean spark-ignition combustion-An experimental study. *Energy*, *191*, p. 116520. https://doi.org/10.1016/j.energy.2019.116520
51. O'Donnell, P. C., J. Gandolfo, B. Gainey, E. Vorwerk, R. Prucka, Z. Filipi, B. Lawler, Hessel, R., Kokjohn, S., Huo, M. and Salvi, A. (2022). *Effects of Port Angle on Scavenging of an Opposed Piston Two-stroke Engine. No. 2022-01-0590.* SAE Technical Paper. https://doi.org/10.4271/2022-01-0590
52. Lin, Y. (2023). Prediction of temperature distribution on piston crown surface of dual-fuel engines via a hybrid neural network. *Applied Thermal Engineering*, *218*, p. 119269. https://doi.org/10.1016/j.applthermaleng.2022.119269
53. Jia, X., J. Willard, A. Karpatne, J. S. Read, J. A. Zwart, M. Steinbach, and V. Kumar. (2021). Physics-guided machine learning for scientific discovery: an application in simulating lake temperature profiles. *ACM/IMS Transactions on Data Science*, *2*(3), pp. 1–26. https://doi.org/10.1145/3447814

4 Fault Detection and Tolerance in Dynamic Control

Enhancing System Reliability and Performance through Dynamic Fault Detection and Tolerance Mechanisms

P. Shudhi Rishaa, C. Shri Raghavi, M. P. Anbarasi, and J. Niresh

4.1 INTRODUCTION

The incorporation of both humans and robots in various dynamic control systems has increased in popularity in the last decade. These human–machine interactions, which range from industrial robotics to autonomous vehicles and medical gadgets, hold enormous promise for improving productivity, efficiency, and safety. However, as these systems become more sophisticated and complicated, they become more prone to errors and failures. To maintain the dependability, safety, and efficacy of human–machine interaction, understanding, recognizing, and mitigating these flaws is crucial. The imperative necessity to overcome the difficulties brought on by discrepancies in dynamic control systems during human-machine interaction serves as the driving force behind this study. Many different types of malfunctions can appear, from sensor issues to communication problems to actuator oddities [1]. These flaws, if ignored, might result in unpredictable behavior, endangering not only the functionality of the system but also the safety of the human operators.

4.1.1 Importance of Fault Detection and Tolerance in Human-Machine Interaction

No amount of emphasis can be placed enough on the need for anomaly identification and tolerance in human–machine interaction. It is crucial for a number of strong reasons, including:

DOI: 10.1201/9781003392330-4

1. **Assurance of Safety:** Defects in dynamic control systems may have serious effects on safety. Critical applications like autonomous vehicles, surgical robots, and industrial automation can be made safer by identifying these errors in real time and putting tolerance measures in place to minimize accidents, injuries, and even fatalities [2].
2. **System Reliability:** In businesses where accuracy and consistency are crucial, reliable human–machine interactions are essential. Systems that are fault tolerant can guarantee continuous operation, minimize downtime, and prevent expensive disruptions.
3. **Human-Centric Design:** When humans interact with machines, they must have faith in and rely on these systems. Providing fault tolerance promotes user trust and automation acceptance [3], allowing for easier transitions in human-centric businesses.
4. **Optimized Performance:** Fault tolerance and detection can also result in increased system effectiveness. System uptime and operational effectiveness can be increased by locating and fixing issues.
5. **Economic Impact:** Dynamic control system failures can have a considerable negative economic impact in terms of maintenance and lost production. These financial risks can be reduced by putting fault tolerance mechanisms in place.

4.2 LITERATURE REVIEW

4.2.1 Overview of Fault Detection and Tolerance Techniques in Dynamic Control

The literature review portion of this research paper offers an in-depth analysis of fault detection and tolerance methods in the context of dynamic control [4]. This includes a look at several approaches and tactics used to find and fix problems in control systems. Important elements of this review include:

1. **Model-Based Approaches:** Analysis of model-based fault detection techniques that use mathematical models of the system to find deviations from predicted performance [5].
2. **Data-Driven Techniques:** An investigation of data-driven fault detection techniques that use sensor data and machine learning techniques to spot anomalies
3. **Sensor Fusion:** Discussion of methods for fusing data from various sensors to more accurately detect faults [6], known as sensor fusion.
4. **Fault Tolerance Strategies:** An examination of fault tolerance mechanisms, such as redundancy, fault isolation, and reconfiguration methods.
5. **Real-time Implementation:** Information on the difficulties and factors to take into account when integrating fault tolerance and detection into dynamic control systems.

4.2.2 Previous Research in Human–Machine Interaction and Control

It provides a solid foundation for comprehending the historical setting and cutting-edge research in this interdisciplinary field [7]. Important components include:

1. **Human-Centered Approaches:** An overview of studies emphasizing user experience and human elements in human–machine interactions is provided by the term "human-centered approaches." [8]
2. **Control Methods:** A look at the several human–machine systems that employ control methods like adaptive control, haptic feedback, and shared control.
3. **Safety and Reliability:** Exploration of earlier research on safety and dependability issues in human-machine interaction, including case studies and takeaways.
4. **Applications:** Showcasing a variety of HMI applications, from teleoperation and industrial automation to robotic surgery and driverless cars.

4.2.3 Relevant MATLAB® Tools and Libraries

These tools are necessary for the conception, implementation, and assessment of fault tolerance, and detection techniques include the following:

1. **MATLAB® Simulink:** A description of the popular software program MATLAB Simulink, which is used to model, simulate, and analyze dynamic control systems [9].
2. **Control System Toolbox:** Discussion of the Control System Toolbox, a collection of features and instruments for control system design and analysis.
3. **Machine Learning and Signal Processing Libraries:** Finding MATLAB libraries for machine learning and signal processing is essential for strategies for data-driven defect identification.
4. **Human-in-the-Loop Simulations:** Consideration of MATLAB's human-in-the-loop simulation capabilities, which allow the evaluation of fault detection and tolerance tactics with user input.

4.3 Dynamic Control in Human–Machine Interaction

In the context of human–machine interaction, dynamic control describes the intricate process by which machines and people interact in real time while adapting to shifting circumstances and user inputs [10]. Continuous modifications are required in this interaction in response to user, environmental, and system needs. In order to facilitate smooth communication and collaboration between humans and machines, dynamic control systems work to maintain stability, precision, and efficiency during these interactions.

4.3.1 Challenges in Maintaining Control and Stability

1. **Human Variability:** People have a wide range of behavior, which makes it difficult to effectively forecast and react to their inputs. Adaptive control algorithms are needed because gestures, preferences, and decision-making are variable.
2. **Real-time Feedback:** Dynamic interactions necessitate real-time feedback and responses [11]. Particularly in applications like virtual reality or robotic systems, delays or lags in communication can ruin the user experience and cause instability.
3. **Environmental Factors:** Environmental elements can interfere with sensors and lower the quality of input data. Examples include noise, poor lighting, and physical obstacles. Dynamic control systems must eliminate background noise and concentrate on pertinent data.
4. **Safety Issues:** It is crucial to ensure the safety of both users and machinery. Dynamic control systems must foresee potential hazards, react quickly to unanticipated events, and avert potentially hazardous circumstances.
5. **Adaptability:** Human intents and preferences can quickly alter due to human adaptability. Dynamic control systems must be adaptable, taking user behavior into account when making changes to their responses to suit changing requirements and preferences.
6. **Complexity of Interactions:** Complex tasks requiring fine control might arise during human-machine interactions, such as in surgical robotics or cutting-edge gaming systems. It is quite difficult to manage the complexity of these interactions and keep everything stable.
7. **Integration of Multimodal Inputs:** In contemporary human–machine interfaces, various input modalities, including touch, voice, and gestures, are frequently used [12]. The seamless integration and appropriate interpretation of these inputs raise the bar for dynamic control's complexity.

In conclusion, dynamic control in human–machine interaction is a complex problem requiring real-time processing power, sophisticated algorithms, and a thorough knowledge of both human behavior and machine dynamics [13]. For intuitive, effective, and secure interactions between humans and machines in a variety of applications, these problems must be overcome.

4.4 FAULT TYPES IN HUMAN–MACHINE INTERACTION

4.4.1 Classification of Faults

Faults in human–machine interaction can be broadly classified into several categories [15], including the following:

1. **Sensor Faults:** These happen when the sensors that the system uses to gather data malfunction, giving incorrect or no data. Sensor errors may cause the system's replies to be misinterpreted as user intents.

2. **Actuator Faults:** Actuator faults are problems with the parts that execute commands from a control system. A system's reactions may be delayed, executed incorrectly, or fail entirely due to malfunctioning actuators.
3. **Communication Errors:** Errors in data transfer between components result in communication faults. This can affect the cooperation between people and machines by causing delays, losing important information, or causing miscommunication.
4. **Faults in the Software:** Defects or faults in the software that manages the interaction process may result in unexpected behaviors, system crashes, or inaccurate interpretations of user input.

4.4.2 Impact on Control Performance and Safety

1. **Control Performance Degradation:** Errors can seriously reduce the effectiveness of the control system, causing sluggish movements, delayed reactions, or wrong actions. Applications needing real-time and precise control, such as autonomous vehicles or surgical robots, can be particularly vulnerable to this degradation [14].
2. **Safety Risks:** Defects in how humans and machines interact can be extremely dangerous. For instance, a sensor flaw in a medical equipment could result in the patient receiving the incorrect dosage, harming them. Similar issues with actuators might impair an autonomous vehicle's reaction time to obstacles, increasing the risk of accidents.
3. **Loss of User Confidence:** Repeated errors can reduce user confidence in the system. Users' reluctance or unwillingness to use the technology could have an effect on its acceptance and adoption.
4. **Increased Workload:** Users may be required to make up for system flaws when they occur, adding to their mental and physical burden. The user experience and general effectiveness may be impacted by this additional burden.

Strong fault detection and tolerance methods are needed to address these flaws, guaranteeing that the system can recognize abnormalities, adjust to changes, and sustain secure and effective human–machine interactions even in the presence of flaws.

4.5 FAULT DETECTION METHODS

4.5.1 Model-Based Fault Detection

Model-based defect detection uses mathematical representations of the behavior of the system. Departures can be recognized as errors by comparing actual system data with the expected behavior predicted by the model [16]. These models might be as intricate as complicated simulations or as simple as analytical formulae. For systems where a thorough grasp of the underlying physics is available, model-based techniques are effective. They require accurate modeling for efficient defect identification since they are sensitive to model mistakes.

4.5.1.1 MATLAB Implementation

When putting into practice a model-based fault detection system, deviations brought on by faults are first detected using a mathematical model of the system under normal operating conditions. Here is an illustration of a straightforward MATLAB model-based fault detection method. In this example, then demonstrate how to detect a sensor malfunction using a linear system as the model (Figure 4.1).

Machine learning-based solutions for data-driven defect identification use algorithms like neural networks, support vector machines, and decision trees to identify trends in past or present data. These algorithms analyze data without using explicit models to identify abnormalities or departures from typical behavior. Data-driven approaches are especially useful for complicated systems when it is difficult to fully model the entire system [17]. They are flexible for a variety of applications because they can adjust to changing circumstances and learn from new information.

4.5.2 Data-Driven Fault Detection

Machine learning-based solutions for data-driven defect identification use algorithms like neural networks, support vector machines, and decision trees to identify trends in past or present data. These algorithms analyze data without using explicit models to identify abnormalities or departures from typical behavior. Data-driven approaches are especially useful for complicated systems when it is difficult to fully model the entire system [17]. They are flexible for a variety of applications because they can adjust to changing circumstances and learn from new information.

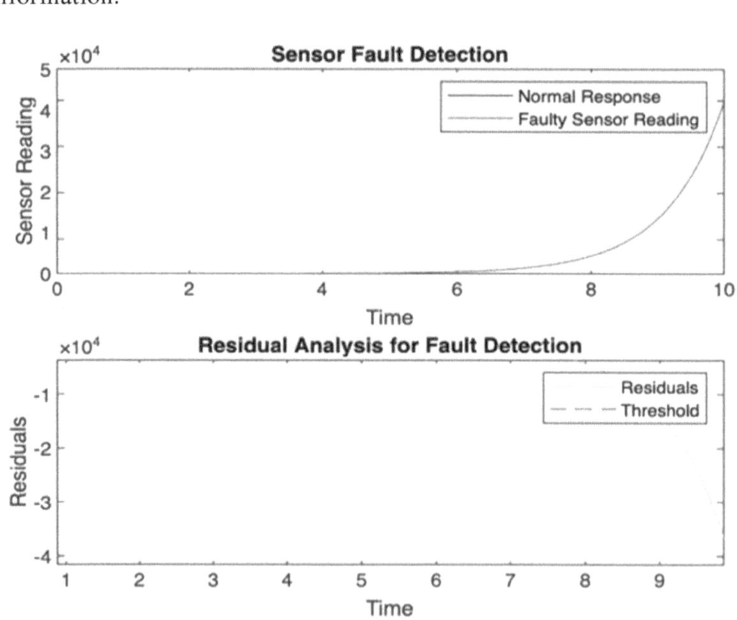

FIGURE 4.1 MATLAB® simulation of sensor fault detection.

4.5.2.1 MATLAB Implementation

Here is a complete MATLAB implementation of an isolation forest-based data-driven fault detection system (Figure 4.2).

The delivered algorithm uses the isolation forest technique to achieve data-driven fault detection. The `fitensemble` function is used to train an isolation forest model after loading data from a CSV file. Every data point's anomaly score is predicted by the model, and anomalies are categorized based on a threshold. Plotting normal data points in blue and discovered anomalies in red indicates the presence of abnormalities above the threshold. This method provides a simple and efficient way to find anomalies in multidimensional data, allowing possible problems or inconsistencies to be quickly identified. The visualization facilitates the comprehension and analysis of the data by offering a clear depiction of abnormalities that have been found.

4.5.3 Sensor Fusion Techniques

In order to provide a more complete and accurate picture of the system, sensor fusion techniques combine data from many sensors. Redundant data can be removed and a more accurate picture of the system's state can be produced by integrating data from many sensors. By enhancing the accuracy and dependability of the input data, sensor fusion improves defect detection. For sensor fusion, methods like Kalman filtering or Bayesian inference are frequently employed. These techniques are essential for applications involving noisy, insufficient, or error-prone sensor data.

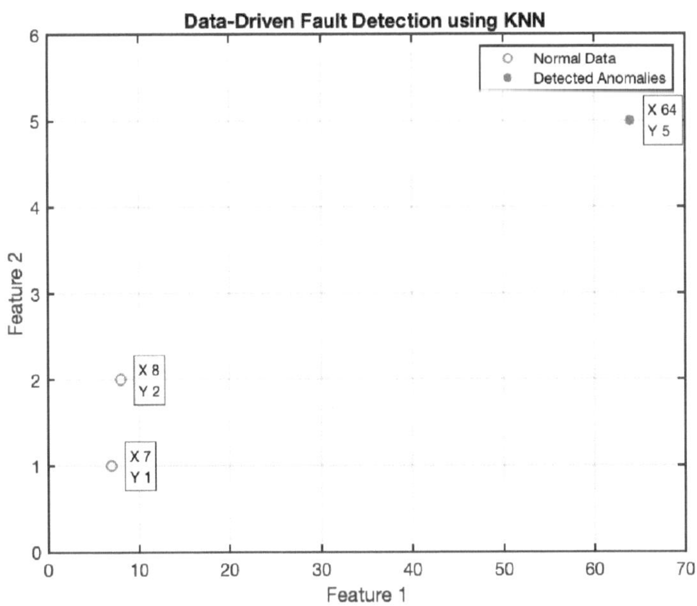

FIGURE 4.2 MATLAB implementation of an isolation forest-based data-driven fault detection system.

4.5.3.1 MATLAB Implementation

In order to improve the precision and dependability of the defect detection process, sensor fusion techniques are used to combine data from many sensors. Using fusion algorithms like Kalman filters or sensor weighting techniques is one typical strategy. The MATLAB implementation of a fundamental sensor fusion technique, specifically a straightforward averaging approach, for defect detection is demonstrated here. The data from three sensors (Sensor A, Sensor B, and Sensor C) must be stored in distinct arrays in order for this example to work (Figure 4.3).

By averaging input from three sensors and then detecting problems based on a predetermined threshold, the output above illustrates sensor fusion. Plotting the fusion result across time allows one to see the combined sensor data. The fusion result is obtained by averaging sensor values. Furthermore, anomalies surpassing the designated threshold are detected using a fault detection technique and shown as binary fault signals. This method provides a simple yet efficient way to combine sensor data and quickly spot possible anomalies or problems in the system. Timely interventions and system monitoring are made easier by the visualization, which helps with the clear interpretation of the fault detection results and the fused sensor data.

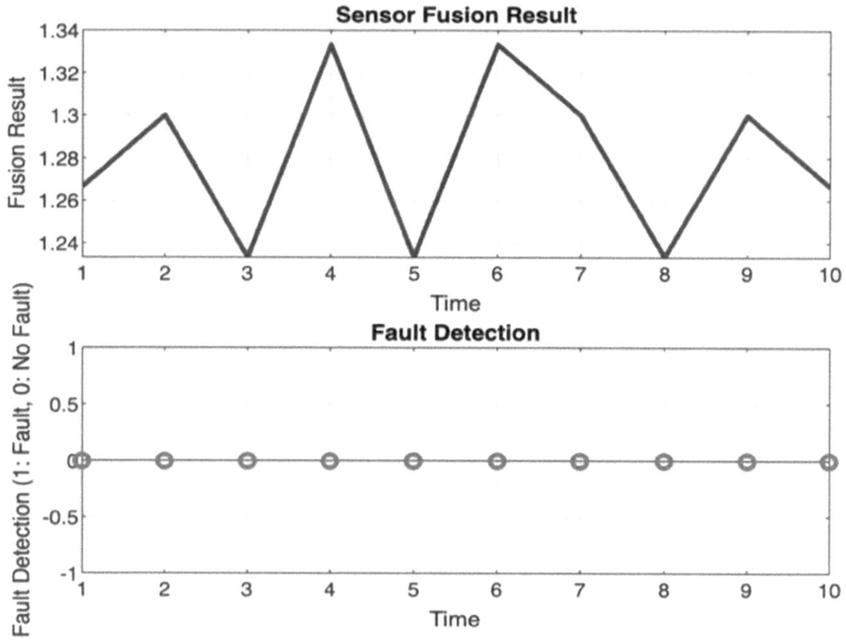

FIGURE 4.3 MATLAB implementation of a fundamental sensor fusion technique.

4.6 FAULT TOLERANCE STRATEGIES

The continued and reliable operation of complex systems, especially in the face of faults or failures, depends on fault tolerance strategies. To sustain system functionality, a number of crucial techniques are used:

4.6.1 Fault Isolation and Identification

- **Definition:** Finding the origin and location of a failure within a system is the goal of fault isolation. The process of classifying the kind of defect that has occurred is known as fault identification.
- **Importance:** Finding the precise defect is essential because it enables targeted repair actions, which reduce downtime and guarantee that resources are used effectively.
- **Methods:** To isolate problems and precisely ascertain their nature, a variety of approaches including sensor data analysis, diagnostic algorithms, and system monitoring are used.

4.6.1.1 MATLAB Implementation

Below is a basic example of how to implement a simple fault isolation and identification algorithm in MATLAB using statistical methods. In this example, it has to assume that sensor data are stored in a matrix called sensor data, where each row represents a different set of sensor readings (Figure 4.4).

The presented application uses both mean and standard deviation criteria to perform fault detection on sensor data. It creates synthetic sensor data, determines

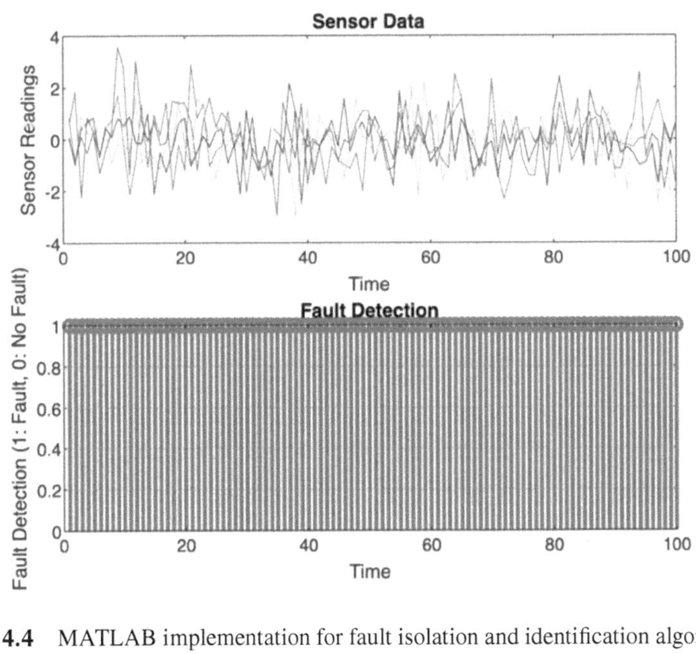

FIGURE 4.4 MATLAB implementation for fault isolation and identification algorithm.

each sensor's mean and standard deviation, and establishes fault detection criteria. To create the final fault identification, the mean-based and standard deviation-based defects are recognized independently and then combined. The graph highlights the sensor readings over time and highlights any defects that have been found. This method makes it possible to monitor and identify changes in sensor behavior in real time, which is essential for preserving system performance and dependability.

4.6.2 Reconfiguration of Control Strategies

- **Definition:** Reconfiguration is the capacity of a system to dynamically modify its control tactics in response to identified defects.
- **Importance:** By modifying control techniques, the system can continue to operate satisfactorily despite errors, avoiding catastrophic failures and preserving stability.
- **Methods:** To adjust control methods based on real-time data and fault information, model-based control, adaptive control algorithms, and artificial intelligence approaches are used.

4.6.2.1 MATLAB Implementation

This MATLAB program mimics changing a first-order system's control techniques and charts the system's response over time. In this illustration, the system output is maintained at a desired setpoint using an adaptive PID controller. The system output and the control signal are then plotted over time by the code for display (Figure 4.5).

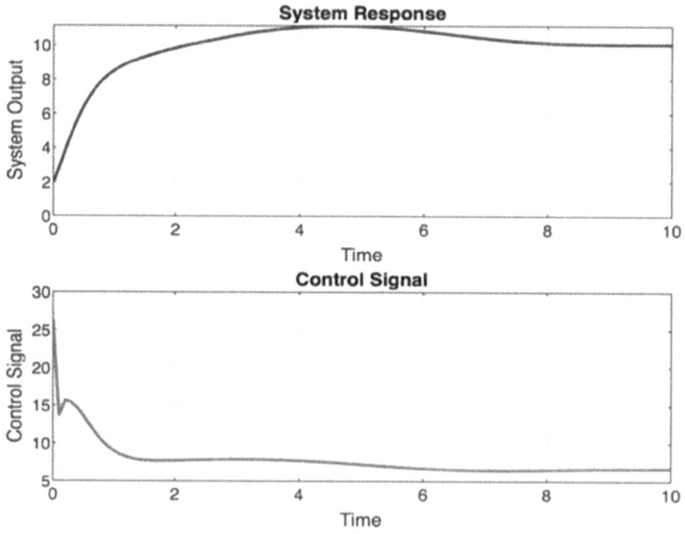

FIGURE 4.5 System's response for changing a first-order system's control techniques.

4.6.3 Redundancy and Backup Systems

- **Definition:** Redundancy involves the replication of critical components or systems within a larger system. Backup systems are secondary systems that can seamlessly take over when the primary system fails.
- **Importance:** Redundancy ensures that even if a component fails, the system can continue operating using backup resources, enhancing reliability and minimizing the impact of failures.
- **Methods:** Redundancy can be achieved through hardware redundancy (duplication of components) or software redundancy (duplication of functions). Backup systems are often designed to be activated automatically when a failure is detected.

4.6.3.1 MATLAB Implementation

- systemAOutput, systemBOutput, and backupSystemOutput are arrays that store the simulated outputs of System A, System B, and the backup system, respectively.
- The simulation loop runs through the time array and calculates the outputs for each system based on their frequencies.
- Fault detection conditions (check for System A failure and check for System B failure) should be replaced with your specific fault detection logic.
- The code then plots the outputs of System A, System B, and the backup system using different colors for differentiation (Figure 4.6).

```
System A has failed. Switching to System B.
System B has failed.
Switching to Backup System.
```

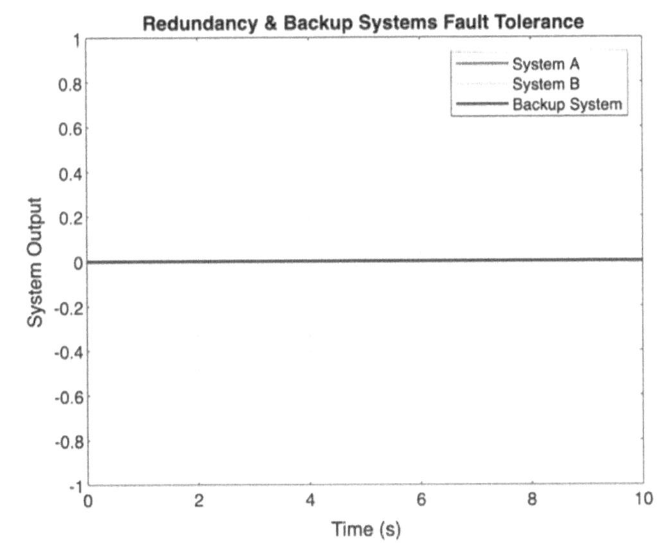

FIGURE 4.6 MATLAB simulation of redundancy and backup systems fault tolerance.

4.7 HUMAN–MACHINE INTERFACE AND INTERACTION

4.7.1 Human Factors in Fault Detection and Tolerance

The design and functionality of fault detection and tolerance systems in human–machine interaction heavily depend on human variables. It is crucial to comprehend how people view, understand, and react to flaws. The capacity to identify and address problems as soon as they arise is influenced by factors like human attention, cognitive load, and situational awareness. When creating fault detection interfaces, engineers must take these aspects into account. To guarantee that crucial information is properly communicated without distracting the operator, alarms, warnings, and notifications should be designed in accordance with human cognitive processes. To improve the defect detection abilities and make sure it can successfully navigate through complicated control interfaces, operators must receive proper training and education.

4.7.2 User-Centered Design Considerations

User-centered design principles are fundamental for creating fault detection and tolerance systems that are intuitive and user-friendly. The interface should be designed with the end-user in mind, considering their expertise, knowledge, and workload. It should accommodate both novice and expert operators. Clear and concise visualization of system status, alarms, and fault information is essential. Information should be presented in a manner that allows for quick comprehension. Feedback mechanisms, such as haptic feedback or auditory cues, can enhance the user's understanding of fault conditions and assist in decision-making. Usability testing and user feedback should be integrated into the design process to identify usability issues and refine the interface.

4.8 INTEGRATION WITH HUMAN FEEDBACK DURING FAULT DETECTION AND TOLERANCE

There are several issues and challenges faced during the process of fault detection and tolerance considering the context of history. One of the main challenges yet to be resolved is integration with human feedback. In human–machine interaction scenarios, the human operator provides constructive feedback, either consciously or subconsciously, which plays a very vital role in fault detection. Integrating this feedback effectively into the detection system without overwhelming it with irrelevant data is a challenge. However, integrating this feedback effectively poses a challenge, and the reasons are:

1. **Relevance:** Not all human commentary is pertinent for fault finding. The system must be able to discern between feedback that suggests a problem or anomaly and purposeful human input (which is a normal element of the control process). False positives may result if the system is overly sensitive and misinterprets natural human input variances as errors. On the other side, if it lacks sufficient sensitivity, it can overlook real flaws.
2. **Variability:** Depending on elements like the operator's expertise, weariness, or emotional state, the human response can be quite variable. It is

difficult to create algorithms that can take this heterogeneity into account while yet identifying recurrent patterns suggestive of errors.

3. **Real-Time Processing:** In applications where split-second choices are critical, integrating human feedback in real time calls for advanced algorithms and quick processing. Accidents or inefficiency could result from slow or sluggish replies.
4. **Filtering and Interpretation:** Human feedback can take on a variety of shapes, including pressure, verbal cues, and physical sensations. This variety of feedback must be appropriately interpreted by the system. Creating algorithms that can distinguish between feedback that genuinely indicates a problem and that is irrelevant or inconsistent is a difficult task.
5. **User Experience:** Including human input shouldn't have a negative impact on the user experience. The human operator may feel uncomfortable or under stress if the technology is too intrusive or demanding, which could result in poor performance or unhappiness.

4.9 EXPERIMENTAL SETUP AND MATLAB IMPLEMENTATION TO OVERCOME THE CHALLENGE

In the experimental setup, there is a dynamic manipulator (such as a robotic arm) designed for human interaction. The manipulator is equipped with force sensors and accelerometers to capture real-time data during interactions. The force sensors measure the forces exerted on the manipulator, while the accelerometers record the acceleration along different axes.

1. **Data Acquisition:**

   Force Sensors: Utilize force sensors and accelerometers to capture human feedback during interactions with the dynamic manipulator. The manipulator is equipped with force sensors that measure the strength and direction of forces applied during interactions. These sensors deliver force information at a predetermined sampling rate.

   Accelerometers: To measure the manipulator's acceleration along the X, Y, and Z axes, accelerometers are strategically positioned on the device. Data from accelerometers are captured at the same rate as that of force sensors.
2. **Data Processing and Pre-processing:**

   Sampling Rate Adjustment: The raw data collected from force sensors and accelerometers might have different sampling rates. To ensure consistency and facilitate analysis, the data are resampled to a target sampling frequency. For example, it resampled the data to 200 Hz.

   Noise Removal and Filtering: Raw force data are filtered using a low-pass filter to remove high-frequency noise. Filtering is essential to ensure that the data used for analysis are smooth and accurate.

   Numerical Differentiation: Accelerometer data are numerically differentiated to calculate acceleration values from raw position data. This step provides acceleration profiles for further analysis.

3. **Fault Detection & Tolerance:**

   Threshold-Based Fault Detection: A threshold-based approach is employed to detect faults in the force data. If the filtered force exceeds a predefined threshold, it is considered a fault. Detected faults are stored with timestamps for further analysis.
4. **Visualization and Analysis:**

   Data Visualization: The filtered force data are visualized using MATLAB, allowing for a clear representation of the force signals and the detected faults. The plot displays the force signal in blue, with detected faults marked in red for easy identification.

   Log and Analysis: The code logs the timestamps of detected faults, allowing for in-depth analysis of the faults concerning the corresponding human–machine interactions. These data can be further analyzed to improve the manipulator's design and fault detection algorithms.

4.10 RESULT AND ANALYSIS

The generated output of the experiment (Figure 4.7):

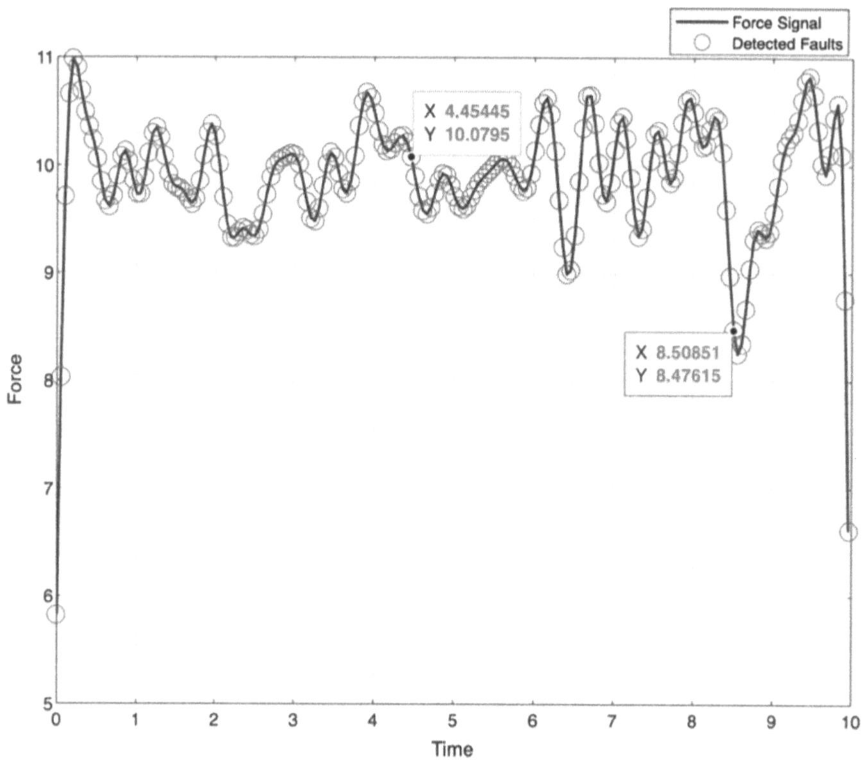

FIGURE 4.7 MATLAB output of the experiment.

Command window:

```
Fault detected at time: 0.000000 seconds
Fault detected at time: 0.050050 seconds
Fault detected at time: 0.100100 seconds
Fault detected at time: 0.150150 seconds
Fault detected at time: 0.200200 seconds
Fault detected at time: 0.250250 seconds
Fault detected at time: 0.300300 seconds
Fault detected at time: 0.350350 seconds
Fault detected at time: 0.400400 seconds
Fault detected at time: 0.450450 seconds
Fault detected at time: 0.500501 seconds
Fault detected at time: 0.550551 seconds

Fault detected at time: 0.600601 seconds
Fault detected at time: 0.650651 seconds
Fault detected at time: 0.700701 seconds
Fault detected at time: 0.750751 seconds
Fault detected at time: 0.800801 seconds
Fault detected at time: 0.850851 seconds
Fault detected at time: 0.900901 seconds
Fault detected at time: 0.950951 seconds
Fault detected at time: 1.001001 seconds
```

Presenting experimental data effectively is crucial for conveying meaningful insights.

Force Data Visualization:

Line Plot: Plot the filtered force data over time to show how they vary during the interaction. Use different colors for different interactions if there are multiple sets of data (Figure 4.8).

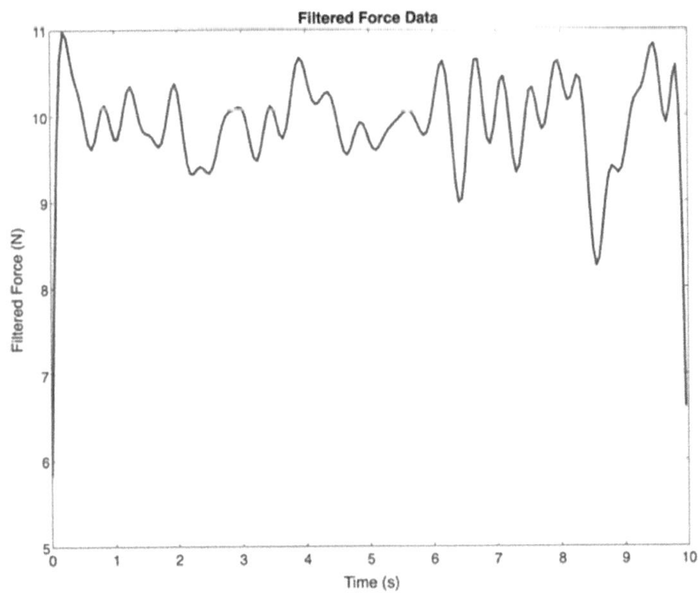

FIGURE 4.8 Force data visualization graph.

Acceleration Data Visualization:

Subplots: Create subplots for each axis (*X*, *Y*, *Z*) to show how the accelerations change (Figure 4.9).

Fault Detection Visualization:

Overlay Plot: Overlay the detected faults on the force plot to visualize when faults occurred (Figure 4.10).

Additional Insights:

Histograms: Show the distribution of forces to understand the range and frequency of different force levels (Figure 4.11).

4.10.1 Interpretation of Results

1. **Analysis of Force Data: Force Distribution**

   The histogram of the filtered force data reveals that the majority of interactions take place within a particular force range, such as 8–12 N. This shows that the force applied to the manipulator is largely constant during ordinary interactions.

   Faults Found:

   The detected fault overlay plot on the force data shows specific time intervals where the force was greater than the threshold (for instance, 15 N). These examples show possible aberrant interactions or circumstances where the manipulator was subjected to exceptionally strong forces.

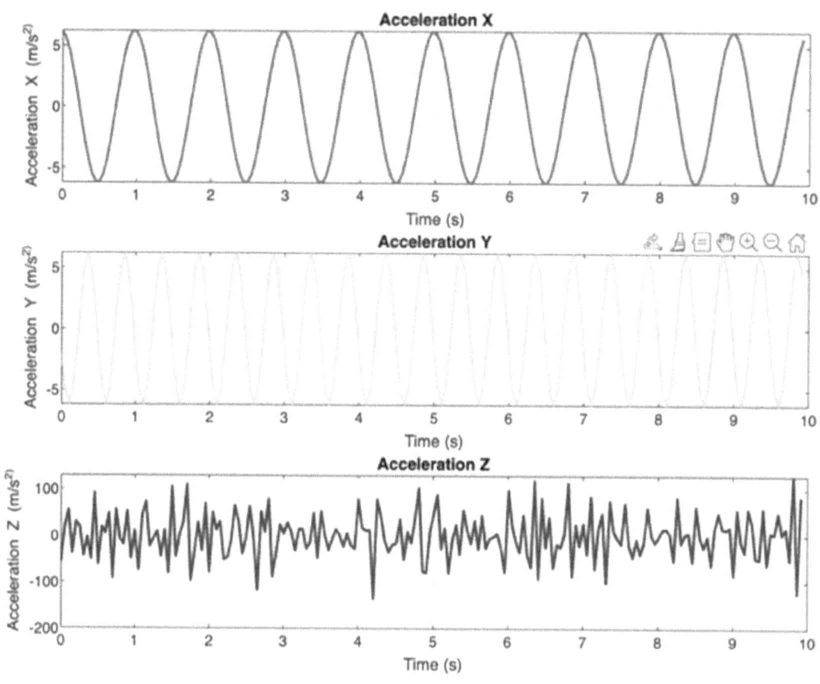

FIGURE 4.9 Acceleration data visualization graph.

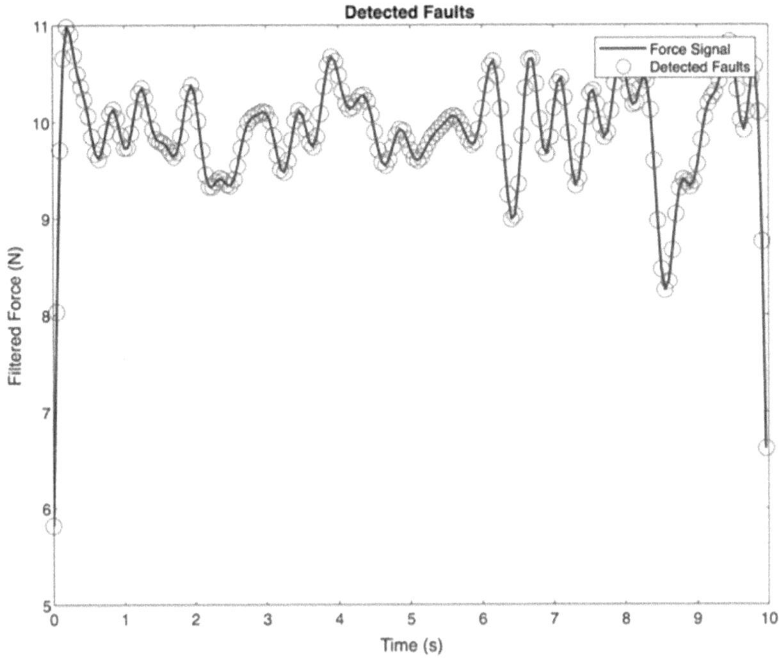

FIGURE 4.10 Fault detection visualization graph.

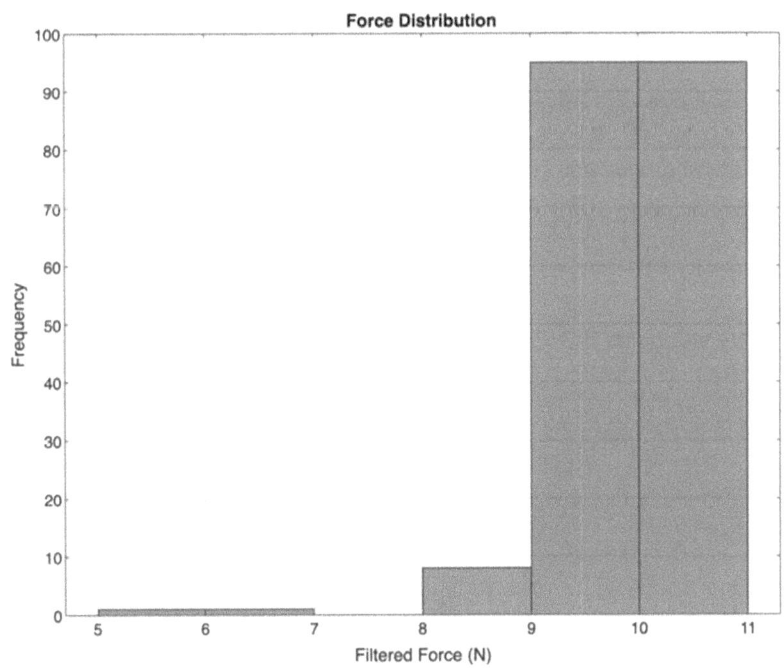

FIGURE 4.11 Histogram for force distribution

2. **Analysis of Acceleration Data**:
   Profiles of Acceleration:
   How the manipulator's acceleration varies over time may be seen in the subplots that display acceleration along several axes. During interactions, sudden spikes in acceleration may be related to hasty movements or abrupt stops.
3. **Overall Interpretation:**
   Normal Interactions:
   Most of the interactions fall within a certain force range, indicating typical user behavior during interactions with the manipulator. These interactions are within the expected force limits and can be considered normal.
   Faulty Interactions:
   Detected faults, indicated by instances where the force exceeded the threshold, represent abnormal or potentially dangerous interactions. Further analysis is needed to understand the cause of these faults. It could be due to user error, technical issues, or unexpected external factors.
   Correlation Insights:
   Analyzing correlations between force and acceleration can provide insights into the relationship between user movements and the forces applied. Positive correlations might suggest that specific manipulator movements lead to higher forces, indicating areas where user training or system improvement might be needed.

4.11 FUTURE IMPROVEMENTS THAT CAN BE IMPLEMENTED

1. **Dynamic Thresholding:**
   Implement adaptive algorithms that dynamically adjust the fault detection threshold based on historical data or user-specific profiles. Machine learning models can learn from past interactions to set optimal thresholds.
2. **Multimodal Sensor Integration:**
   Integrate additional sensors such as vision systems or pressure sensors on the manipulator. Combining data from multiple sensors enables a more nuanced understanding of interactions, enhancing fault detection accuracy.

4.12 CONCLUSION

In this study, it is been explored the intricate challenges and innovative solutions surrounding fault detection and tolerance in dynamic control systems during human–machine interaction. Through a comprehensive review of existing literature, it also delved into various fault detection techniques, ranging from model-based approaches to data-driven methods and sensor fusion techniques. The integration of human feedback in the fault detection process emerged as a critical aspect, bridging the gap between automated algorithms and human decision-making processes. The findings of this study hold significant practical implications for fields such as robotics, manufacturing, and assistive technology. Enhanced human–machine interactions can lead to safer and more productive workplaces, improved manufacturing processes, and advanced assistive devices for individuals with mobility impairments.

In conclusion, the integration of force sensors and accelerometers has proven to be invaluable for understanding and improving human–machine interactions with dynamic manipulators. As technology continues to evolve, these advancements pave the way for a future where machines respond seamlessly to human inputs, ultimately transforming various sectors and enhancing the quality of human–machine collaboration. With ongoing innovation and collaboration between researchers, engineers, and end-users, the possibilities for creating intelligent, adaptive, and user-friendly machines are boundless.

REFERENCES

1. M. L. Visinsky, I. D. Walker and J. R.Cavallaro, (1994). "New Dynamic Model-Based Fault Detection Thresholds for Robot Manipulators." In *Proceedings of the 1994 IEEE International Conference on Robotics and Automation*. IEEE.
2. J. Viana and K. Cohen. (2019). "Fault Tolerance Tool for Human and Machine Interaction & Application to Civilian Aircraft." In *2019 IEEE Latin American Conference on Computational Intelligence (LA-CCI)*, USA (pp. 1–2). IEEE.
3. Madhavi & Kaur. (2017). "Trust-Based Fault Tolerance in Mobile Ad-Hoc Networks Using Adaptive Monitoring." In *2017 International Conference on Computing, Communication and Automation (ICCCA)*, Ecuador.
4. R. Ramezani & Y. Sedaghat. (*2013*). "An overview of fault tolerance techniques for real-time operating systems." *ICCKE*, 2013, pp. 1–6.
5. S. X. Ding. (2008). "Integrated Design of Fault Detection Systems." In *Model-Based Fault Diagnosis Techniques* (pp. 369–401). Berlin, Heidelberg.
6. J. W. C. Van Lint & S. P. Hoogendoorn. (2009). "A Robust and Efficient Method for Fusing Heterogeneous Data from Traffic Sensors on Freeways." *Computer-Aided Civil and Infrastructure Engineering*, 25(8), pp. 596–612.
7. Z. Du & N. Lyu. (2023), *The Influence of Human-Machine Interaction Information on Takeover Behaviors During Human-Machine Co-Driving Control Transition*. Elsevier BV, Amsterdam.
8. J. Woo and N. Kubota (2017). "Human Machine Interaction and Robotics." In *2017 10th International Conference on Human System Interactions (HSI), Ulsan, South Korea, July 17–19, 2017.*
9. MathWorks. (n.d.). *Simulink - Simulation and Model-Based Design – MATLAB – MathWorks*. https://in.mathworks.com/products/simulink.html
10. Y. Wang & F. Zhang. (2017). *Trends in Control and Decision-Making for Human-Rrobot Collaboration Systems*. Springer International Publishing, Cham.
11. M. Tiator, F. Büntig, & C. Geiger. (2018). "Dynamic Movement Monitoring – Algorithms for Real Time Exercise Movement Feedback." In *Proceedings of the 4th International Conference on Information and Communication Technologies for Ageing Well and e-Health, Funchal, 22.03. 2018–23.03. 2018* (pp. 184–191). SCITEPRESS-Science and Technology Publications.
12. P. M. Jones & C. M. Mitchell. (1991). "Human-Machine Cooperative Interaction in the Supervisory Control of a Complex Dynamic System." In *Conference Proceedings 1991 IEEE International Conference on Systems, Man, and Cybernetics* (pp. 1301–1306). IEEE.
13. D. Orellana & A. Madni. (2012). "Extending Model Based Systems Engineering for Human Machine Interaction Analysis and Fault Tolerant Design". In *Infotech@ Aerospace 2012* (p. 2537). American Institute of Aeronautics and Astronautics (AIAA), Reston, VA.

14. S. Liu. (2020). "Chassis Technologies for Autonomous Robots and Vehicles." In *Engineering Autonomous Vehicles and Robots: The DragonFly Modular-based Approach* (pp. 23–34). IEEE
15. M. Mansouri, M. F. Harkat, H. N. Nounou, & M. N. Nounou. (2020)–"Model-Based Approaches for Fault Detection". In *Data-Driven and Model-Based Methods for Fault Detection and Diagnosis.* Springer, Cham.
16. Z. Chen. (2020). "The Basics of Fault Detection." In *Data-Driven Fault Detection for Industrial Processes.* Springer Vieweg, Wiesbaden.
17. G. Antonelli. (2003). "A Survey of Fault Detection/Tolerance Strategies for AUVs and ROVs." In Caccavale, F., Villani, L. (eds), *Springer Tracts in Advanced Robotics.* Fault Diagnosis and Fault Tolerance for Mechatronic Systems: Recent Advances Springer, Berlin.

5 Connected Cars, Connected Cities
The Rise of Smart Mobility

N. Murugu Nachippan and Balaji Vasudevan

5.1 INTRODUCTION TO SMART MOBILITY

Smart mobility is a multimodal approach to transportation that integrates various modes of transportation into a seamless network. It is powered by real-time data and intelligent systems, allowing for seamless connections and optimizing travel times. The foundation of smart mobility lies in two key pillars: technological advancements, such as IoT sensors, data analytics, cloud computing, and artificial intelligence (AI), and collaboration and open data [1].

Smart mobility isn't just about making cars smarter; it's about rethinking the entire transportation ecosystem. It promotes a shift towards a multimodal approach, integrating various modes of transportation – public transit, cycling, walking, and car-sharing – into a seamless network [2].

The chapter explores the rapidly evolving world of connected and autonomous vehicles, providing insights into the future of mobility and the role that technology will play in shaping transportation systems.

Smart mobility is a key component in smart city initiatives that are currently being explored around the world by authorities, industry players, and academics alike as discussed by the authors, which encompasses many facets of mobility, including improving public transport services, providing guidance to commuters and motorists, and real-time traffic monitoring and management, among others [3].

Figure 5.1 presents a web of interconnected concepts in the realm of smart transportation and urban development. The central theme revolves around how modern technologies and infrastructures integrate to create smarter, more sustainable cities. Key elements like Telematics & Data Sharing and 5G Network Integration enable communication technologies such as vehicle-to-vehicle communication and vehicle-to-infrastructure communication, which are essential for autonomous driving and connected cars. These advancements support Smart Traffic Management and Smart Parking Solutions, which are critical for efficient urban mobility. The Internet of Things and Smart Mobility Ecosystem serve as foundational technologies, linking various systems like electric vehicle charging networks, transportation integration, and connected cities. Sustainability and green energy are also integral, contributing to the push for eco-friendly urban planning and smart grid systems. Together, the interconnected concepts suggest a future where cities use integrated technologies to enhance mobility,

DOI: 10.1201/9781003392330-5

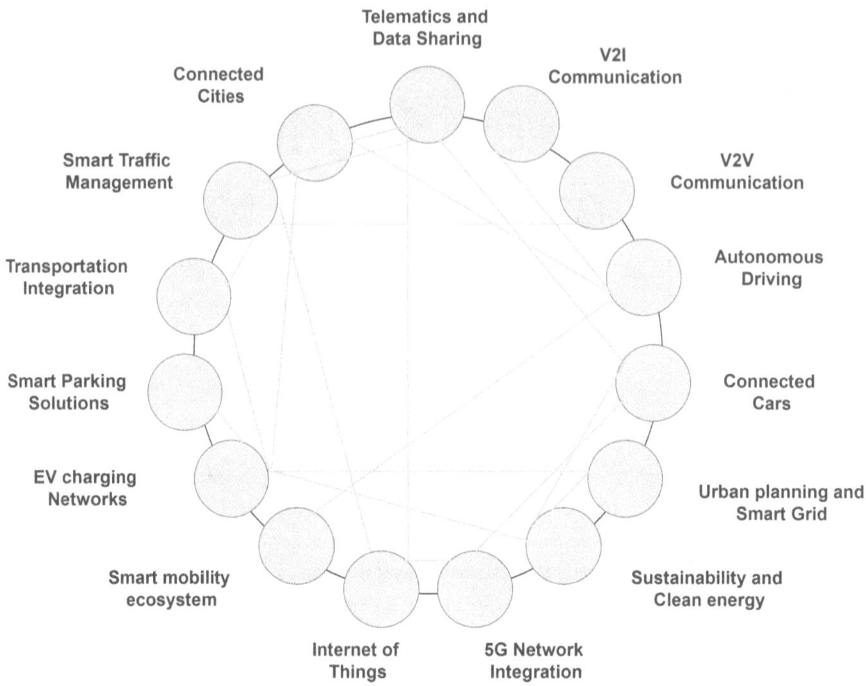

FIGURE 5.1 Interconnected framework of smart mobility and urban infrastructure ecosystem.

reduce environmental impact, and optimize resources across urban landscapes. This conveys how all these elements are linked, working together to create a seamless, efficient, and sustainable transportation ecosystem within smart cities.

5.2 TECHNOLOGICAL INNOVATIONS DRIVING SMART MOBILITY

Technological advancements involve IoT sensors collecting real-time data on traffic flow, parking availability, and environmental conditions, which are then processed by AI algorithms to optimize traffic flow, predict congestion, and suggest alternative routes. Collaboration and open data platforms foster innovation and the development of innovative mobility solutions [4].

A well-implemented smart mobility system offers numerous benefits, including reduced traffic congestion, improved air quality, enhanced public safety, economic growth, and increased accessibility for people with disabilities [5]. However, challenges include significant infrastructure investment, public acceptance and privacy concerns, and integration and standardization. The world of transportation is undergoing a revolution driven by advancements in vehicle technology, with electric vehicles (EVs) and autonomous vehicles (AVs) leading the charge [6]. EVs are becoming a popular choice for eco-conscious drivers due to battery breakthroughs, faster charging infrastructure, superior torque and instant acceleration, and near-zero tailpipe emissions. EVs also contribute to cleaner air and combat climate change [7]. AVs promise a future where cars drive themselves, transforming commutes and unlocking new possibilities. They rely on a complex suite of sensors like LiDAR, radar, and cameras to perceive their surroundings, and AI to make critical decisions like steering, braking, and lane changes. High-definition maps and localization technologies are crucial for AVs to understand their position and navigate precisely [8].

Figure 5.2 illustrates key technological innovations driving smart mobility, organized into five interconnected categories. Autonomous vehicles rely on sensors, cameras, AI, and V2X communication to enable self-driving technology and improve road safety. EVs are advancing with better battery technology, expanded charging infrastructure, and integration of renewable energy, promoting sustainability. Connected infrastructure, powered by IoT and 5G networks, facilitates smart traffic management and real-time data sharing between vehicles and urban systems. Mobility-as-a-Service focuses on shared transportation through ride-sharing platforms, integrated payment systems, and data analytics to optimize urban travel. Lastly, sustainable urban mobility emphasizes modernizing public transportation, promoting electric micro-mobility such as e-scooters and e-bikes, and integrating smart city planning for efficient, eco-friendly mobility solutions. Together, these innovations form the foundation of smarter, more sustainable, and efficient urban transportation ecosystems. The integration of these technologies not only enhances road safety by potentially eliminating human errors that lead to accidents but also holds the promise of reducing traffic congestion, improving travel efficiency, and offering increased mobility to diverse populations, including the elderly and individuals with disabilities. By addressing technical, regulatory, and societal challenges, AVs are poised to revolutionize commuting experiences and unlock new realms of possibilities in the realm of transportation.

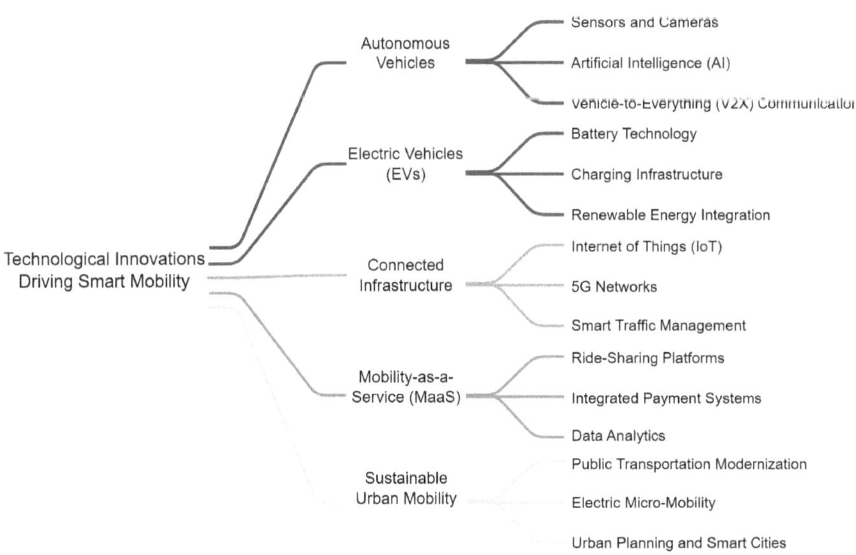

FIGURE 5.2 Key technological innovations shaping the future of smart mobility.

However, both EVs and AVs face challenges, such as cost barriers for EVs and the need for regulatory frameworks to govern their operation on public roads. Despite these challenges, the potential of both EVs and AVs is undeniable. As technology evolves, we can expect more affordable EVs, more ubiquitous charging infrastructure, gradual advancement in AVs, and collaboration between car manufacturers, technology companies, and policymakers [9]. EVs face cost barriers due to battery technology limitations, while AVs require regulatory frameworks. Future advancements may lead to affordable EVs, widespread charging infrastructure, and improved AV technology through collaboration [10].

The **Internet of Things (IoT)** is revolutionizing the way vehicles interact with the world, transforming them from isolated machines into intelligent, connected companions. IoT in automobiles involves embedded sensors that monitor various aspects of a vehicle's performance, environment, and driver behaviour. These data are then transmitted wirelessly to a cloud platform, where they are analysed and used to generate insights and enable communication. The Internet of Things (IoT) enhances vehicles by integrating sensors for performance monitoring, environmental data, and driver behaviour analysis, transmitting to the cloud for insights and communication [11]. Automotive IoT utilizes embedded sensors in vehicles to monitor performance, environment, and driver behaviour [12]. Data are wirelessly sent to the cloud for analysis, enhancing vehicle intelligence and connectivity [13].

Vehicle-to-Everything (V2X) communication allows vehicles to exchange information with each other and roadside infrastructure, enabling car-to-car, car-to-infrastructure, and vehicle-to-pedestrian communication. This enhances safety, improves traffic flow, personalizes driving experiences, and allows for proactive maintenance. The connected car ecosystem opens doors for innovative mobility services, such as car-sharing platforms and on-demand ride-hailing. The proposed antenna enhances V2X applications with an extended wireless communication range. Antenna covers 0.75 GHz to 7.6 GHz with 164% bandwidth, efficiency around 90%, and longer wireless communication range for V2X [14]. RetroV2X meets V2X needs in various scenarios effectively. Mechanisms handle collisions and improve performance at high vehicle densities [15].

However, **connected car technology** faces challenges such as security concerns, standardization, and infrastructure development. Robust cybersecurity measures are essential to protect sensitive vehicle information. A lack of uniform standards across different manufacturers can hinder seamless communication and integration of systems. Collaborative efforts are needed to establish common protocols and ensure the full potential of V2X communication is realized. In this study, the authors explored an evaluation model for smart transportation innovation policies, including connected and automated vehicles (CAVs), smart traffic safety, transportation management systems, intelligent-based transportation technology, transportation resource integration and sharing, and traffic data collection. The study utilized the analytic hierarchy process (AHP) and decision-making trial and evaluation laboratory (DEMATEL) to analyse the weighted values, correlation, and degree of influence between different dimensions and criteria of smart transportation innovation policies.

5.3 INTEGRATION OF ARTIFICIAL INTELLIGENCE AND MACHINE LEARNING

Artificial intelligence (AI) and machine learning (ML) are driving the connected car revolution, transforming cars from passive machines into intelligent transportation marvels. AI includes advanced driver-assistance systems (ADAS), navigation and route optimization, personalized driver experiences, and predictive maintenance. Benefits of AI-powered cars include enhanced safety, improved efficiency, and greater comfort and convenience [16].

AI algorithms are revolutionizing the automotive industry across multiple fronts. In advanced driver-assistance systems (ADAS), AI processes data from sensors like cameras and radar to detect potential hazards such as pedestrians or obstacles on the road. This technology enables ADAS to take pre-emptive actions like automatic braking or steering adjustments, significantly enhancing vehicle safety. Moreover, AI-driven navigation systems utilize real-time traffic data and weather conditions to recommend optimal routes, ensuring drivers reach their destinations efficiently and safely. Beyond safety, AI personalizes the driving experience by adjusting climate control, seat positions, and entertainment settings based on individual preferences, creating a more comfortable and tailored environment for each driver [17].

Additionally, AI's role extends to predictive maintenance, where it analyses vehicle data to forecast potential issues before they manifest. By monitoring engine performance, tire conditions, and other metrics, AI can alert drivers and service providers to impending maintenance needs, thereby reducing downtime and extending the longevity of vehicles. This proactive approach not only enhances reliability but also contributes to cost savings and customer satisfaction. As AI continues to evolve, its integration into automotive technology promises further advancements in safety, efficiency, and personalized driving experiences, marking a transformative era in the automotive industry.

ML, a subset of AI, plays a crucial role in data analysis, continuous improvement, and personalized experiences in connected cars. Advantages include enhanced safety features, optimized performance, and more precise maintenance predictions.

Figure 5.3 illustrates how AI and ML technologies drive smart mobility outcomes through their integration with connected cars and connected cities. AI and ML technologies provide data processing and infrastructure analysis to connected cars, enabling autonomous driving and traffic prediction. Connected cars, in turn, contribute data that help connected cities optimize traffic management and public transport systems. Connected cities use these data to improve urban mobility and provide feedback to AI and ML systems for continuous improvement. The collaboration between these technologies and systems ultimately leads to enhanced smart mobility outcomes, such as efficient traffic management and optimized public transport.

5.4 INFRASTRUCTURE AND URBAN PLANNING FOR SMART MOBILITY

Infrastructure investment is crucial for implementing smart mobility solutions, while public acceptance and privacy concerns arise from the integration of technology.

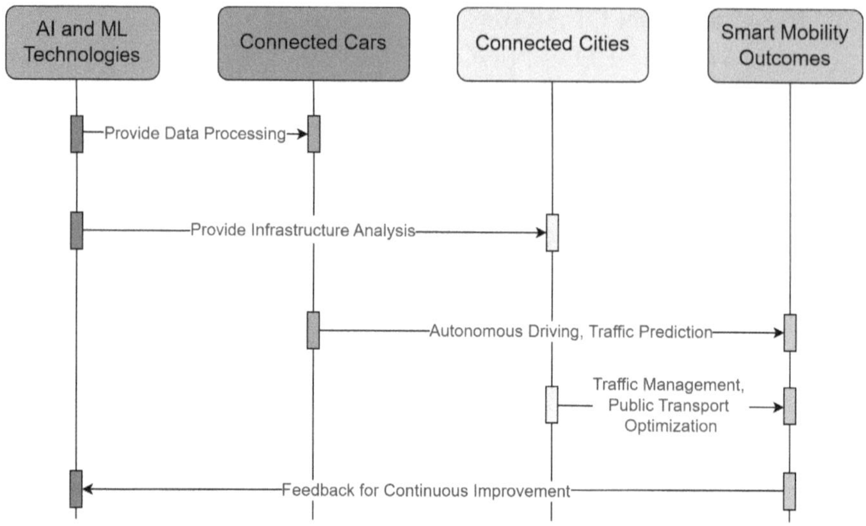

FIGURE 5.3 Integrating AI, connected cars, and cities for enhanced smart mobility solutions.

Collaboration and standardized protocols are essential for integrating different transportation systems and ensuring data compatibility across platforms. Overall, smart mobility aims to make transportation more accessible, convenient, and affordable for everyone.

5.4.1 Smart Traffic Management Systems (STMS)

The brains behind the operation, STMS, utilize a network of sensors embedded in roads and traffic lights. These sensors collect real-time data on traffic flow, congestion points, and even weather conditions. These data are then fed into AI algorithms that optimize traffic light timings, display dynamic route suggestions for drivers, and provide real-time information to travellers. Imagine traffic lights that adjust based on congestion, reducing wait times and easing bottlenecks [18].

Promoting a shift towards sustainable modes of transportation like cycling, walking, and micro-mobility requires dedicated infrastructure. Protected bike lanes, physically separated from vehicle traffic, create a safe environment for cyclists, encouraging more people to choose this healthy and eco-friendly option. Wide sidewalks and pedestrian plazas prioritize pedestrians by providing larger walkways, pedestrian-only zones, and shared spaces, fostering a more walkable and vibrant urban setting. Additionally, dedicated bus lanes give priority to public transit, allowing buses to avoid congestion, maintain schedules, and offer a more efficient and attractive alternative to car travel. Together, these measures enhance urban mobility and contribute to a more sustainable:

Electric Vehicle Charging Infrastructure: A network of conveniently located charging stations is crucial for widespread EV adoption. This includes fast-charging stations for long-distance travel and slower chargers in residential areas and public parking spaces.

Connected and Autonomous Vehicle (CAV) Integration: While fully autonomous vehicles are still under development, infrastructure needs to be future-proofed to accommodate them. This may involve dedicated lanes for CAVs, communication systems embedded in roads, and digital signage for real-time information exchange.

Smart city transportation design goes beyond just infrastructure; it's about creating a city layout that fosters a multimodal approach. Here's how urban planning plays a crucial role:

Mixed-Use Development: Imagine living, working, shopping, and enjoying entertainment all within walking or cycling distance. Mixed-use development reduces reliance on cars for everyday errands, promoting a more walkable and sustainable lifestyle.

Transit-Oriented Development (TOD): Developing high-density residential and commercial areas around public transport hubs encourages ridership. This creates a virtuous cycle where people have easy access to public transport, making car ownership less essential [19]. Figure 5.4 gives the strategic infrastructure investment for autonomous mobility and smart city integration.

Compact Urban Design: Densely packed, walkable neighbourhoods reduce the need for long commutes. This not only improves convenience but also fosters a stronger sense of community.

Pedestrian-Friendly Design: Prioritizing pedestrians means incorporating elements like wider sidewalks, street trees for shade, crosswalks with proper signage, and accessible pedestrian signals. These features create a more inviting and comfortable environment for walking.

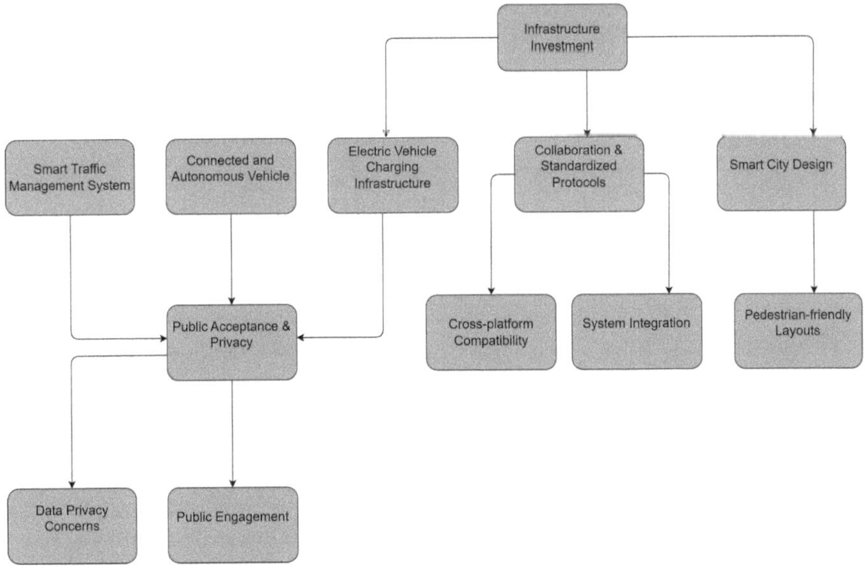

FIGURE 5.4 Strategic infrastructure investment for autonomous mobility and smart city integration.

Greener Cities: Urban green spaces like parks and street trees not only improve air quality but also encourage walking and cycling. Additionally, strategically placed trees can provide shade and reduce the heat island effect in cities.

5.5 SHARED MOBILITY SERVICES

It is revolutionizing transportation, offering a convenient and affordable alternative to car ownership. Key players include ridesharing, bike-sharing, and scooter-sharing, which connect riders with drivers through mobile apps. Benefits of shared mobility include reduced traffic congestion, improved accessibility, and first- and last-mile connectivity.

Public transit systems, such as bus rapid transit (BRT) and light rail and subway expansions, are essential for efficient urban transportation networks. BRT systems use dedicated lanes, rapid boarding systems, and priority at intersections to reduce travel times and attract more riders. Light rail and subway networks provide high-capacity options for longer distances, reducing reliance on cars. Real-time information and integration through apps and digital displays improve user experience and encourage ridership [20].

The Smart Mobility project as discussed by the authors aims at designing measures to encourage the increased use of public and non-motorized transport by integrating behavioural economic principles into public policy, and the extensive involvement of citizens and their participation in the design of the measures are to support their democratic legitimization and later acceptance. Last-mile solutions, such as micro-mobility options like bike-sharing, scooter-sharing, and dockless e-bike rentals, can bridge the gap between public transport stops and final destinations. On-demand shuttles and walkable urban design can also help encourage people to use public transport or park-and-ride options. Overall, shared mobility services are transforming the way we travel and provide a more sustainable and convenient alternative to car ownership [21].

5.6 DATA ANALYTICS AND URBAN MOBILITY MANAGEMENT

5.6.1 Predictive Analytics: Predicting the Future of Traffic

Big data is not just about collecting information; it's about using it to predict future trends and patterns. Predictive analytics are crucial in transportation planning, including traffic congestion prediction, public transit demand forecasting, and proactive maintenance [22]. By analysing historical data and real-time sensor information, algorithms can identify congestion hotspots, suggest route adjustments, optimize scheduling and resource allocation, and predict potential maintenance issues, enabling preventive measures and avoiding disruptions. Figure 5.5 illustrates how data collection and analytics drive urban mobility management. Data are gathered from traffic sensors, public transit systems, mobile apps, GPS, and environmental sensors. These data undergo analysis to enable real-time traffic analysis, predictive modelling, and route optimization. These processes inform smart traffic signals and dynamic route guidance, improving public transport optimization and resource allocation. Additionally, demand forecasting supports urban mobility management by providing valuable insights for policy and planning, enhancing overall urban transportation systems.

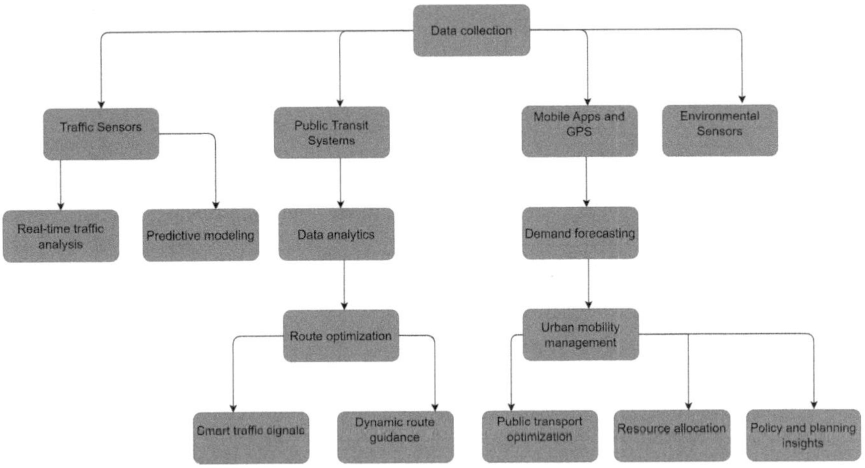

FIGURE 5.5 Data-driven urban mobility management for optimizing traffic and public transport systems.

5.6.2 Data-Driven Decisions for Optimized Networks

Big data provides valuable insights for transportation planners, enabling them to make data-driven decisions that optimize transportation networks. These include dynamic traffic management, where real-time data can adjust traffic timings, deploy variable speed limits, and reroute traffic based on congestion. Public transit network optimization involves identifying underutilized routes and high-demand areas, allowing for route adjustments and targeted service improvements [23]. Infrastructure investment planning also benefits from data analysis, identifying areas with high traffic volume, frequent accidents, or deteriorating infrastructure.

5.6.3 Conclusion

Smart mobility represents a paradigm shift in urban transportation, fuelled by technological innovations such as AI, ML, and advanced data analytics. These technologies play pivotal roles in optimizing urban mobility management by integrating various elements like infrastructure, urban planning, and shared mobility services. AI and ML algorithms analyse vast amounts of data generated from sensors, traffic cameras, and mobile devices to provide real-time insights into traffic patterns, congestion levels, and user preferences. This data-driven approach enables cities to implement efficient transportation strategies, such as dynamic routing and congestion pricing, to improve traffic flow and reduce environmental impact.

REFERENCES

1. Edith, Maier. "Smart mobility – Encouraging sustainable mobility behaviour by designing and implementing policies with citizen involvement." *JeDEM: eJournal of eDemocracy and Open Government* (2012). doi: 10.29379/JEDEM.V4I1.110

2. Alok, Prakash. "Smart mobility solutions for a smart city." *IEEE Potentials* (2021). doi: 10.1109/MPOT.2020.3023539
3. Thomas, Schulz, Thomas, Schulz, Markus, Böhm, Heiko, Gewald, Helmut, Krcmar. "Smart mobility – an analysis of potential customers' preference structures." *Electronic Markets* (2021). doi: 10.1007/S12525-020-00446-Z
4. Ovidiu, Vermesan, Reiner, John, Patrick, Pype, Gerardo, Daalderop, Kai, Kriegel, Gerhard, Mitic, Vincent, Lorentz, Roy, Bahr, Hans, Erik, Sand, Steffen, Bockrath, Stefan, Waldhör. (2021). Automotive intelligence embedded in electric connected autonomous and shared vehicles technology for sustainable green mobility. doi: 10.3389/FFUTR.2021.688482
5. Luke, Butler, Tan, Yigitcanlar, Alexander, Paz. "How can smart mobility innovations alleviate transportation disadvantage? assembling a conceptual framework through a systematic review." *Applied Sciences* (2020). doi: 10.3390/APP10186306
6. Timo, Birnschein, Christian, Oekermann, Mehmed, Yuksel, Benjamin, Girault, Roman, Szczuka, David, Grünwald, Sven, Kroffke, Mohammed, Ahmed, Yong-Ho, Yoo, Frank, Kirchner. "Enhancing mobility using innovative technologies and highly flexible autonomous vehicles" (2014). doi: 10.1007/978-3-319–08087-1_5
7. Hemant, Nandanpawar. "Electric vehicles for low carbon sustainable development of transport sector of developing Asia." *IRA-International Journal of Technology & Engineering* (2017). doi: 10.21013/JTE.ICSESD201735
8. Aboelmagd, Noureldin, Mohamed, Elhabiby. "A framework for multi-sensor positioning and mapping for autonomous vehicles." *The International Archives of the Photogrammetry, Remote Sensing and Spatial Information Sciences* (2023). doi: 10.5194/isprs-archives-xlviii-1-w1-2023-339-2023
9. Dave, Butler, Jörn, Mehnen. "Challenges for the adoption of electric vehicles in thailand: potential impacts, barriers, and public policy recommendations." *Sustainability* (2023). doi: 10.3390/su15129470
10. Sachin, Baliram, Shahapure, Vandana, Kulkarni, Sanjay, M, Shinde. "A technology review of energy storage systems, battery charging methods and market analysis of EV based on electric drives." *International Journal of Electrical & Electronics Research* (2022). doi: 10.37391/ijeer.100104
11. P. Raja, Dr.S.Senthil, kumar, Digvijay, Yadav, Dr. Taresh, Singh. "The internet of things (IOT): a review of concepts, technologies, and applications." *International Journal of Information Technology and Computer Engineering* (2023). doi: 10.55529/ijitc.32.21.32
12. Ahmad, Faiz, Ab, Rahman, Hazlina, Selamat, Ahmad, Jais, Alimin, Mohd, Taufiq, Muslim, Muhammad, Mazizan, Msduki, Nurulaqilla, Khamis. "Automotive real-time data acquisition using wi-fi connected embedded system" (2019). doi: 10.11113/ELEKTRIKA.V18N3-2.189
13. Risabh, Mishra, M, Safa, Aditya, Anand. "Internet of vehicles: commencing intellectual hoarse towards self-regulating cars and vehicular clouds for smart transportation structure [Vehicular ad-hoc network: a review and application in the internet of vehicles]." *International Journal of Engineering and Technology* (2018). doi: 10.14419/IJET.V7I3.12.16176
14. Nathan, Seongheon, Jeong, Ehsan, Soleimanian. "A compact ultra-wideband monocone antenna with folded shorting wires for vehicle-to-everything (V2X) applications." *Sensors* (2023). doi: 10.3390/s23136086
15. Chenren, Xu, Kenuo, Xu, Lilei, Feng, Bo, Liang. "RetroV2X: a new vehicle-to-everything (V2X) paradigm with visible light backscatter networking." *Fundamental Research* (2023). doi: 10.1016/j.fmre.2022.01.038
16. Vaishnav, Datey, Sandeep, Prabhu. "Smart traffic control and prediction model empowered with 5G technology", *Artificial Intelligence and Machine Learning* (2023). doi: 10.1109/aikiie60097.2023.10389965

17. Gyugeun, Yoon, Joseph, Y. J. Chow. "A sequential transit network design algorithm with optimal learning under correlated beliefs." arXiv.org, undefined (2023). doi: 10.48550/arXiv.2305.09452
18. Bushra, Zalloom. "Smart cities: using GIS technology in urban infrastructure development at migration areas." *The Eurasia Proceedings of Science Technology Engineering and Mathematics* (2022). doi: 10.55549/epstem.1192335
19. Farish, Jazlan, Ramin, Saedi, Ali, Zockaie, Mehrnaz, Ghamami, Michelle, Boucher, Hediye, Tuydes-Yaman, Mehdi, Ganji, Andrea, Marr. "Smart city: a mobility technology adoption framework incorporating surface-level technical analysis." *Current Urban Studies* (2022). doi: 10.4236/cus.2022.103023
20. Christos, Gkartzonikas, Loukas, Dimitriou. "Shared micro-mobility services for university communities: a multivariate ordered probit approach." *Transportation Research Record* (2023). doi: 10.1177/03611981231164383
21. Max. T., Ng, Hani S., Mahmassani, Omer, Verbas, Taner, Cokyasar, Roman, Engelhardt. "Redesigning large-scale multimodal transit networks with shared autonomous mobility services." arXiv.org (2023). doi: 10.48550/arxiv.2307.16075
22. Sanika, Nitin, Kunjir, Bhakti, Sunil, Hingane, Janvi, Anand, Pagariya, Mamoon, Rashid. "Managing smart urban transportation with the integration of big data analytic platform." (2023). doi: 10.1109/ic3i59117.2023.10397915
23. Diego, Altafini, Federico, Mara, Valerio, Cutini. "A data-driven approach for a city-university mobility plan: the case of the University of Pisa." *Lecture Notes in Computer Science* (2023). doi: 10.1007/978-3-031-37126-4_27

6 Analysis of Cognitive Internet of Vehicles and Its Challenges

S. Umamaheswari and G. Gowtham

6.1 INTRODUCTION

The origins of concepts like the Internet of Things (IoT) and cognitive computing can be traced back to the 1950s, albeit under different terms. The increasing utilization of IoT technologies is a result of technological progress that has rendered integrated circuits and sensors more accessible, cost-effective, and energy efficient. These advancements have concurrently facilitated the seamless occurrence of machine-to-machine (M2M) communications [1]. The term "Internet of Things" (IoT) describes a broad category of objects that are equipped with digital sensors that are globally networked through the Internet or other technologies. The number of linked devices was estimated to be over 13 billion in 2015. By 2020, it is expected to have increased to over 38 billion, a stunning increase of more than 285%. A wide range of interconnected devices is included in this massive network, including cars, appliances, lighting, drones, food containers, electric metres, manufacturing systems, buildings, infrastructure (bridges, tunnels), wearable technology, cameras, security systems, and many more [2].

Since the 1970s, there has been a notable increase in the worldwide circulation of automobiles, making them the principal mode of daily mobility for human beings. However, despite efforts to reduce traffic mishaps from the point of origin to the point of resolution, there is still a high frequency of these incidents because of things like poor visibility, tired drivers, speeding, and other associated problems. Research indicates that human error or poor judgement accounts for 90% of road accidents [3]. Presently, the automotive industry is undergoing a substantial technological transformation to address the aforementioned challenges. This transformation has been accelerated since 2012, driven by the swift advancement in the IoT [4].

The Internet of Vehicles (IoV) has become the key technology that will enable the implementation of future autonomous driving scenarios. All IoT systems have unique characteristics, and the paradigm underlying those characteristics is based on cognition and autonomicity [5], and consequently, for the IoV as well. According to a report [6], it is projected that by the year 2030, fully autonomous vehicles will make up 15% of the worldwide vehicle market, and these vehicles will be outfitted with both intelligence and communication capabilities.

 DOI: 10.1201/9781003392330-6

This chapter is divided into seven sections. In Section 6.2, an introduction provides the definition of IoT and Cognitive IoT (CIoT). Section 6.3 focuses on the Cognitive Internet of Vehicles (CIoV, covering its evolution employed in CIoV. Section 6.4 discusses the overview, architecture of CIoV and its security challenges, and communication technologies employed in CIoV. Section 6.5 explores the integration of edge computing and 5G in CIoV. Future research directions are explored in Section 6.6. Finally, Section 6.7 serves as the conclusion, summarizing the key findings and the essence of the entire chapter.

6.2 INTERNET OF THINGS AND COGNITIVE INTERNET OF THINGS

6.2.1 Internet of Things

The IoT is a system utilizing unique system identifiers (UIDs) to establish connections among computer devices, mechanical and digital machines, objects, or individuals. This facilitates the transmission of data across a network without necessitating direct human-to-human or computer-to-human interaction. The process involves integrating physical devices with sensors, microprocessors, and communication hardware to gather and share data, enabling the remote monitoring and control of diverse devices and systems. IoT facilitates the seamless exchange of information between the physical and digital realms, creating a network of interconnected devices and enabling automation, data analysis, and enhanced decision-making [7].

According to Cisco's calculations, the inception of the IoT occurred between 2008 and 2009, marked by the surpassing of global population figures by the number of connected objects. By 2010, the count of these objects had nearly doubled, reaching 12.5 billion. Subsequent to this, IoT has become pervasive in everyday life, driven by continuous technological progress and substantial corporate investments [8]. As per IoT analytics projections, the current global tally of connected things stands at 20 billion, contributing to a market valued at $150 billion within the IoT industry. By 2024, there will be more than 30 billion linked objects, with a market worth of roughly $1 billion. IoT may have three types of hurdles to overcome, as with any new technological trend: business, society, and technology [9,10].

6.2.2 Cognitive Internet of Things

The concept of "Cognitive Internet of Things" (CIoT) entails merging cognitive computing technologies with the IoT. Within a CIoT framework, the devices linked to the IoT network go beyond functioning solely as sensors and actuators; they also exhibit the capability to comprehend, analyse, and learn from the data they generate and receive. Figure 6.1 illustrates the fundamental features of CIoT.

Because of intelligent automation, predictive analytics, and proactive intervention offered by IoT technology, the next generation of smart buildings, cars, and manufacturing applications is being created. Complex IoT platforms include powerful machine learning (ML) or artificial intelligence (AI) in various IoT architecture stages. Neural networks, deep learning, natural language processing (NLP),

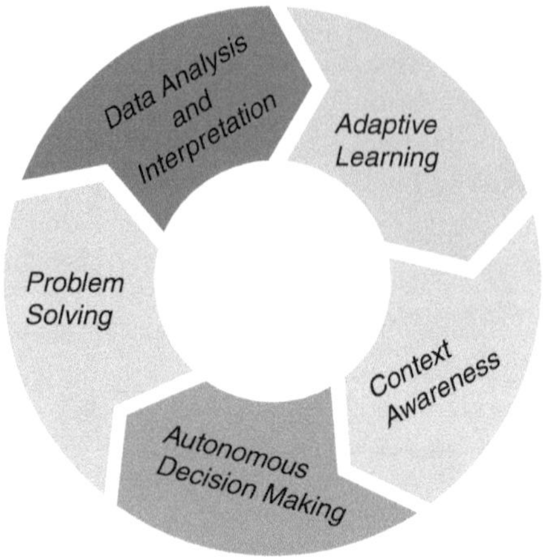

FIGURE 6.1 Key concepts of CIoT.

and other technologies and methodologies make up these components [11]. Using these techniques, autonomous IoT systems that can understand, learn, predict, adapt, and operate on their own are created, going beyond the limitations of traditional rule-based algorithms. These systems can be used to create a wide range of intelligent implementations, including consumer electronics, robotics, autonomous vehicles, and application and service creation [12].

Currently, the IoT is in its early stages, mainly focused on connecting physical objects for communication. However, to unlock the full potential of IoT, it's crucial to enhance the intelligence of these objects. Cognitive computing is playing a significant role in achieving machine intelligence within IoT. This means making objects smarter. To do this, edge intelligence becomes essential. This approach aims to enable efficient communication, fast computing, and intelligent control within the IoT system [13]. According to Ref. [14], CIoT has the potential to enhance the scalability, adaptability, and interactivity of IoT systems in building applications. Additionally, it aids in addressing challenges in various sectors such as residences, smart urban environments, and Industry 4.0.

6.2.2.1 Architectural Layers of CIoT

Research is actively exploring the integration of cognitive elements into the IoT framework, with a primary emphasis on meeting user's requirements for IoT services that possess autonomous and intelligent features. The architecture of autonomously CIoT consists of a three-layer cognitive cycle structure, mechanisms for sharing information, and interactions with heterogeneous circumstantial sensing [15] (Figure 6.2).

The conventional IoT architecture forms the basis for the framework of the CIoT architecture. The CIoT is characterized by four layers: information sensing layer, network interconnecting layer, cognitive decision-making layer, and intelligent service layer.

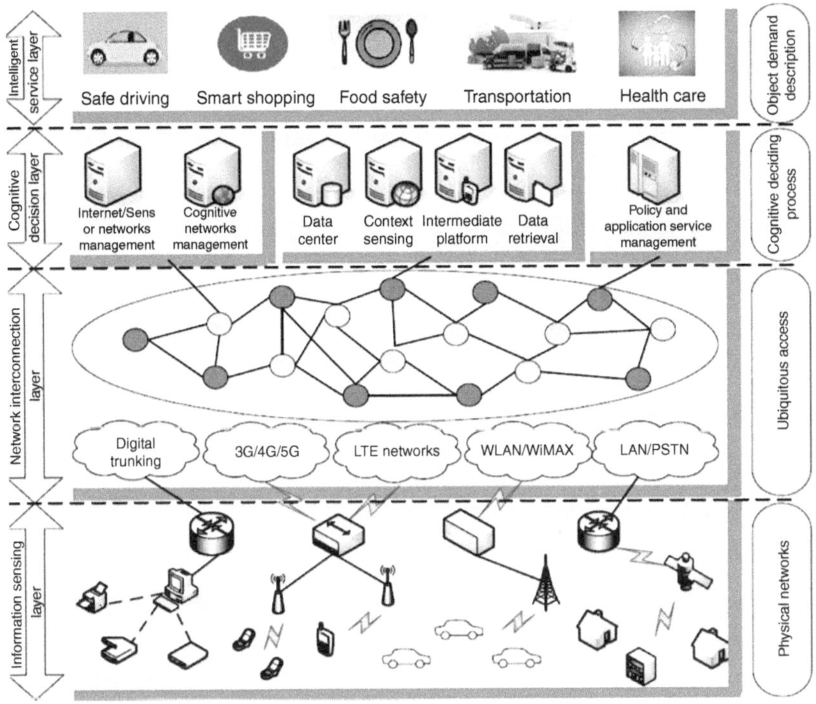

FIGURE 6.2 Architecture of CIoT [16].

The functionality of CIoT can be broken down into four layers: ubiquitous access, cognitive decision-making, object demand description, and physical networks. The four layers' definitions are not exactly matched by the layers of conventional network protocol. The main function involved in this architecture is sensing quality of service (QoS) and achieving network efficiency objectives for users from different cells. Once the network behaviour has been accurately modelled, make the appropriate decision based on feedback, network status, and cognition in self-learning. Finally, determining the CIoT behaviour will be needed in the future, all the while redistributing and modifying physical network resources to satisfy user demand in real time.

The integration of cognitive computing with the IoT enhances the overall intelligence and efficiency of connected systems, enabling a new level of sophistication in data processing, analysis, and decision-making.

6.3 COGNITIVE INTERNET OF VEHICLES

The concept of "Cognitive Internet of Vehicles" denotes a scenario in which vehicles are furnished with cognitive computing capabilities and interconnected seamlessly through the IoT. In this context, cognitive computing involves employing advanced technologies like ML and AI to empower cars with the capacity to sense, comprehend, and make intelligent decisions.

6.3.1 Evolution of Internet of Vehicles

This portion will investigate the cognitive internet of vehicular networks (CIoV) as a progressive approach to augment the cognitive intelligence within the IoV. The segment explores the development of CIoV, emphasizing its differences from intelligent transportation systems (ITS), vehicular ad-hoc networks, and IoV. Furthermore, we conduct an extensive examination of vital technologies essential for bringing CIoV into fruition, including self-driving technology (Figure 6.3).

6.3.1.1 Intelligent Transport System, VANET, IoV

According to Ref. [17], advancements in information and communication technologies (ICT) within the hardware, software, and communications domains have rendered intelligent and sustainable transport systems viable. The incorporation of ICT into the transportation infrastructure enables enhanced and safer travel experiences, facilitating the adoption of ITS. These ITS, grounded in sustainability, integration, security, and responsiveness, will contribute to achieving three fundamental objectives: providing access and mobility, ensuring environmental sustainability, and fostering economic development. Adhering to these principles is imperative for the successful realization of intelligent transportation systems.

In Ref. [18], one of the primary challenges encountered by intelligent transportation systems involves effectively managing the diverse array of sensing devices present in vehicles, on public roads, and within transportation infrastructure. Current challenges for sensing systems include road hazards like potholes, sudden shifts in pavement materials, floods, and adverse environmental conditions that pose risks to vehicle occupants. Additional obstacles comprise the removal or obscuring of transit lines, inadequate or, in certain instances, absent traffic signals, and the swift detection of various objects such as pedestrians, cyclists, debris, tyre residues, or animals. However, determining

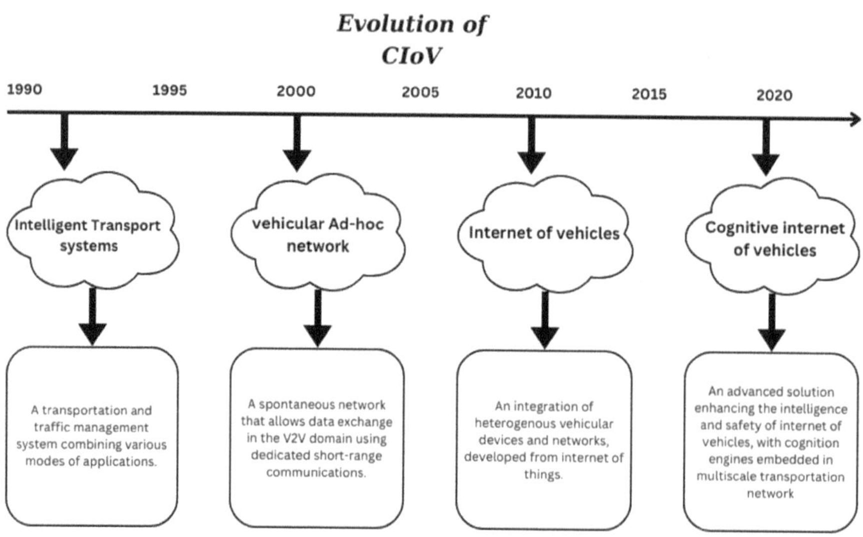

FIGURE 6.3 Evolution of CIoV.

the true boundaries requires extreme caution (such as sensor reachability and the algorithms used to evaluate the sensor data). For instance, we might not receive the proper reaction in time to stop a crash if the front sensors of a moving car identify an object or another car and the algorithms take too long to interpret the data and choose the control system or warning interface. Apart from enhancing vehicle response times across diverse driving scenarios, effective algorithms for integrating data from multiple sources, infrared and photogrammetric systems, as well as various sensor types, could also improve the accuracy of maps. This improvement involves more precisely identifying traffic and potential road hazards, thereby reducing the risk of accidents.

In Ref. [19], ad hoc networks operating within the automotive domain are referred to as VANETs. VANET has become a vital part of the ITS and has emerged as a solution. Enhancing road safety and traffic efficiency are the two main goals of ITS. Through constant internet access, VANET facilitates the transfer of regular data (files, audio, video, etc.) and information about traffic analysis and road safety. VANET is categorized into two main application areas: security applications, which include traffic analysis, interactive driving, and collision avoidance, and user applications, which include internet connectivity on roads, entertainment, and other roadside services like restaurant or fuel pump information.

Information transmitted insecurely over VANET communication could have disastrous consequences. For this reason, the information must be precise, effective, and trustworthy. The goal of every single project in the VANET area is to effectively promote road safety by regularly exchanging data among network nodes. Any successful attack has the potential to cause catastrophic mishaps, fatalities, or financial losses. So, some security pre-requesteries and challenges are listed in Table 6.1.

According to Ref. [20], the term "Internet of Vehicles" (IoV) denotes a platform utilizing diverse communication methods to facilitate information exchange between a vehicle and its surroundings. IoV creates a unified network that performs diverse functions, including intelligent traffic management, dynamic information services, and intelligent vehicle control. This integration occurs through the merging of Internet of Things (IoT) technology with ITS. The IoV comprises three essential elements: intra-vehicular network, vehicular mobile Internet, and inter-vehicular network. This connectivity allows vehicles to remain constantly linked to the Internet,

TABLE 6.1
Security Pre-requesteries and Challenges

Pre-requesteries	Challenges
Authentication	Consistency of data
Reliability	High mobility
integrity	Error tolerance
Anonymity	Latency control
Availability	Key management
Delay handling	
Confidentiality	

forming a network wherein interconnected vehicles share data for services like entertainment, traffic management, and road safety.

The IoV facilitates information exchange among vehicles, road infrastructure, passengers, drivers, sensors, and electric actuators. This communication is made possible by utilizing protocols and standards like IEEE 802.11p, Directional Medium Access Control, Vehicular Cooperative Media Access Control, Ad Hoc On-Demand Distance Vector, Dynamic Source Routing, General Packet Radio Services, and various other specifications.

In Ref. [21], the advancement of the IoV will drive progress in automotive and information technology, fostering social and economic growth. This will result in the creation of a transportation system that is more reliable, efficient, and intelligent, directly influencing the lifestyles of its users. Like other markets, a few trends are driving the IoV market. In this regard, in Ref. [22], the authors noted four key trends that will support the future expansion of IoV:

1. Energy efficiency
2. Connected devices.
3. Security
4. Safety.

6.3.1.1.1 Research Challenges in Internet of Vehicles

IoV implementation is still regarded as a difficult problem. In this section, a few research challenges are covered (Table 6.2).

TABLE 6.2
Summary of Research Challenges in IoV

References	Research Field	Challenges
[23,24]	*Human-vehicle network model and service model*	Research on an effective human-vehicle network in IoV is lacking. To close this gap and improve flexibility to intricate space-time changes, research on service features during coordination and the development of a cognitive learning model are necessary.
[25–27]	*Radio propagation model*	Because information spreads quickly in IoV, designing realistic radio propagation models is essential. WLAN performance is impacted by both dynamic (such as other vehicles) and static (such as buildings) impediments, therefore both need to be taken into consideration.
[28–30]	Localization accuracy	Because of less precise GPS devices and poor signals in crowded cities, accurate car localization on the Internet of Vehicles is difficult. To solve localization concerns, vehicle speed must be taken into consideration.

(*Continued*)

TABLE 6.2 (*Continued*)
Summary of Research Challenges in IoV

References	Research Field	Challenges
[31]	*Enhancing communication ability*	Because of its dynamic nature, IoV requires improved cross-layer communication capabilities. Optimizing users, services, and networks requires addressing issues including bandwidth maintenance during congestion and network traffic reduction.
[31]	*Cooperation technologies of virtual vehicles with drivers*	The collaborative platform of IoV connects drivers to virtual cars via in-car operating systems. Cooperation includes engaging with drivers as well as perceiving and obtaining information. It is essential to design these stages for network performance.
[31]	*Sustainability of service providing in IoV*	IoV encounters difficulties in offering long-term services to different vehicle types (smart, mediocre, and dumb) because of limitations such as scarce bandwidth, intricate urban environments, and fluctuating network circumstances. It is essential to address these problems.
[32–34]	*Location privacy*	Maintaining anonymity in automotive settings is essential to preventing identity theft. Pseudonyms and Mix-Zone are two methods with limitations that make location privacy in IoV difficult to achieve.
[35–38]	*Location verification*	Verifying location data is essential for ensuring passenger safety in automobile communications. This is difficult because there isn't a reliable authority. There are still problems with cooperative approaches and directional antennas.
[39]	*Operational management*	Diverse sensors, radio terminals, transponders, and network operators working together under third-party virtual operators give rise to the complexity of the Internet of vehicles. Managing cloud and fog computing systems presents security and credibility problems.
[40]	*Absence of standardized protocols hindering resilient vehicle-to-vehicle (V2V) communication.*	Open standards are necessary for IoV to provide effective communication and information sharing while improving user experiences. For the ecosystem to work effectively, these standards should guarantee a transparent and smooth connection with the current closed standards.

6.3.1.2 Self-Driving Technology

Artificial intelligence (AI) technology has been gaining momentum in recent years, with deep learning emerging as the key component of AI technology. The automotive sector is very interested in self-driving/autonomous driving technology as a vertical application of AI. According to the Eno Centre for Transportation's investigative report, self-driving technology has the potential to drastically lower the number of traffic accidents brought on by negligent driving [41]. IoV and AI-based self-driving technologies can work well together. On the one hand, a lot of data produced by driving a car can give artificial intelligence (AI) a good foundation for learning and training. However, deep learning algorithms [42] now perform much better in real-time processing due to the rapid development of electronic circuits such as GPUs, TPUs, FPGAs, ASICs, and others. This guarantees immediate business confidence for environmental perception, control, and decision-making in CIoV. Substantial advancements have been made in self-driving research, introducing algorithms like the AI-driven path optimization algorithm [43] and the algorithm for detecting obstacles and roads [44].

6.4 OVERVIEW OF COGNITIVE INTERNET OF VEHICLES

Vehicles can do more in a CIoV system than just connect and exchange data. They can analyse and comprehend their immediate surroundings, decipher intricate traffic situations, and make defensible decisions using the information they are given. This covers functions like adaptive cruise control, predictive maintenance, real-time traffic analysis, and improved safety features.

In Ref. [45], the authors discussed that CIoV focuses on the application of cognitive computing, IoT, and fog computing for autonomous driving. It addresses issues like as the poor intelligence of lone autonomous cars, the requirement for ultra-high reliability and ultra-low latency in complicated driving tasks, and the high cost and restricted efficacy of existing sensor-based systems. The proposed solution from the author utilizes AI-driven decision-making, the exchange of information between vehicles, and a versatile computing paradigm comprising the CIoV. This incorporates edge, fog, and cloud computing. Also, the author highlights the need for efficient data source computation, adaptable computing locations, and intelligent decision-making optimization to improve the reliability of autonomous driving systems.

In Ref. [46], the study examines the integration of cognitive intelligence into the IoV, focusing on employing cooperative techniques and cognitive intelligence to enhance data transmission and communication within vehicular ad hoc networks in the realm of the CIoV. It covers cooperative spectrum sensing, the use of cognitive radios, and the difficulties and solutions associated with routing in cognitive radio networks. Enhancing vehicle-to-vehicle communication, dynamic spectrum access, and network performance are the main priorities. This study discusses the benefits of applying cognitive intelligence to this subject and offers predictions for future advancements in vehicle communication and collaboration.

In Ref. [47], the study investigates the performance of CIoV in the presence of both licensed and unlicensed user activities and mobility. It involves the development of formal analytical models for spectrum sensing, focusing on the probability of misdetection and the estimated duration of overlapping time for mobile secondary users. Additionally, the authors outline details such as the anticipated transmission time and the number of resend attempts following unsuccessful packet delivery. Results show that elevated vehicle speeds are associated with a higher probability of misdetection in spectrum sensing and a decreased expected overlapping time duration in each epoch. The study proposes that leveraging opportunistic communications can enhance the performance of networks in CIoV. Financial support for the project was provided by the National Science Foundation (NSF) and the U.S. Department of Homeland Security (DHS).

Regarding the IoV, a novel approach to routing is covered in Ref. [48]. To be more precise, this study presents SDCoR, a software-defined cognitive routing system created to meet the difficulties associated with routing in dynamic vehicle environments. The design, application, and possible advantages of SDCoR in enhancing the effectiveness and dependability of routing on the IoV are probably covered.

In Ref. [49], the study introduces a privacy-aware content caching architecture for CIoV. It proposes that vehicles and roadside units (RSUs) cache contents in advance and provide them through broadcasting to meet the content needs of vehicles. This protects the sensitive information of vehicles and improves content access. The cognitive engine analyses data and sets caching policies, ensuring that content acquisition latency is within the contact time between vehicles and RSUs. The author also discusses the deployment of content caching when vehicles act as content providers and recommends specific contents to vehicles. Overall, it addresses privacy protection and content quality in CIoV.

In Ref. [50], the study presents a method for securing content caching in 5G-enabled CIoV. This approach incorporates a cognitive engine capable of forecasting and storing data through ML or deep learning models, ensuring effective content delivery with a focus on privacy protection. The approach addresses concerns related to the storage capacity of RSUs and vehicles. The author establishes metrics such as cache hit ratio and latency to evaluate content caching performance, underscoring the significance of effective content caching, security, and privacy in the context of CIoV. In Ref. [51], the authors employed game theory and cognitive computing within the IoV framework to outline a coordinated lane change strategy involving multiple vehicles on a roundabout.

6.4.1 Architecture of CIoV and Security Challenges

In the specified architecture [52], Internet-based cognitive intelligence is seamlessly integrated into a comprehensive five-layered transportation framework designed for the future automotive industry. The layers encompass Sensing and Participation, Internet Cloud, Vehicular Cloud, Vehicle-to-Vehicle Communication, and Strategic Services. Emphasizing a meticulous focus on security considerations at each layer, the architecture prioritizes the security and privacy aspects within the IoV. Notably, it underscores the distinction between the Vehicular Cloud and the Internet Cloud, highlighting the latter's limitations and its role in confined data processing and storage, with the former offering benefits to all users. The primary objectives of the architecture centre around real-time services and security considerations tailored specifically for autonomous vehicles (Figure 6.4, Table 6.3).

6.4.2 Communication Technologies in CIoV

The communication technologies employed in the CIoV include various wireless communication technologies to enable seamless interactions between vehicles and the infrastructure. Some key communication technologies in CIoV are shown in Table 6.4:

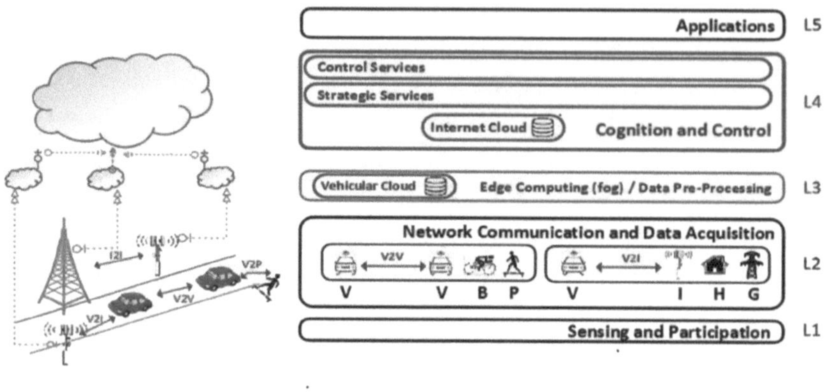

FIGURE 6.4 Architecture of CIoV [52].

TABLE 6.3
CIoV Architecture Layers and Security Challenges

Levels	Name of the Layer	Summary	Security Challenges
Level 1	Sensing and participation	Layer-1 involves smart vehicles and road infrastructure, with autonomous and connected vehicles utilizing automation and wireless communication, enabling AI for improved transportation management in IoV [48].	In Layer-1 of CIoV, autonomous vehicle attacks such as GPS spoofing, map poisoning, and confusing RADAR and LIDAR sensors are security and privacy concerns. Threats to connected cars include tracking, denial-of-service attacks, privacy violations, and data breaches.
Level 2	Network communication and data acquisition	Layer-2 in CIoV oversees network communication for transport data. Intra-vehicular involves smart vehicles' sensor-based internal communication. Inter-vehicular includes V2V and V2I, connecting vehicles, devices, pedestrians, bicycles, and infrastructure for enhanced connectivity [52].	Layer-2 of CIoV is divided into sub-layers V2V and V2I. Problems include message alteration, tampering, tracking, and unauthorized access. Attacks, spoofing, and denial of service are examples of threats [53].
Level 3	Edge computing and data pre-processing	Layer-3 on the vehicular cloud provides edge computing, collecting, and processing data from lower layers for real-time services like navigation and traffic monitoring. It uses fog computing to manage increasing vehicles efficiently [54].	Using intelligent computation from the communications of participating vehicles, this layer offers real-time services. Vehicular cloud security issues include integrity and confidentiality breaches. Real-time services are vulnerable to threats such as hacking and eavesdropping software, data integrity issues, and privacy concerns [55].

(Continued)

TABLE 6.3 (*Continued*)
CIoV Architecture Layers and Security Challenges

Levels	Name of the Layer	Summary	Security Challenges
Level 4	Cognition and Control	The Cognition and Control layer, above the internet cloud, processes lower layer data, providing Control and Strategic services. Strategic services include driver behaviour analysis and road condition monitoring with cloud-based dynamic cognition. The Control sub-layer ensures system performance through security, traffic optimization, and resource allocation using SDN and NFV [56–58].	Security concerns pertaining to the Internet Cloud are addressed in the Cognition and Control layer. Multitenancy vulnerability, unapproved access, availability concerns, abuse, cross-border data flow, trust, and recognized dangers including data loss and service hijacking are among the issues to be concerned about [59].
Level 5	Application layer	The CIoV application layer provides high-level asynchronous functions as end products, supporting Intelligent Transportation System goals for driver assistance and efficient traffic management. It fosters coordination, marking a significant paradigm shift [55].	Delays in data reaching high-level services cause security problems in this layer, which affect objectives like accident reduction. Format incompatibilities cause problems with data interoperability [60].

TABLE 6.4
Summary of Communication Technologies Employed in CIoV

References	Technologies	Summary
[61–65]	V2V communication	M2M communication allows devices and vehicles to exchange information without human intervention. It plays a crucial role in supporting automation and intelligent decision-making in CIoV.
[66,67]	V2I communication	This encompasses the interaction between vehicles and roadside elements like traffic lights, sensors, and other infrastructure units. It enables the exchange of information to enhance traffic optimization, intelligent control of traffic signals, and overall improved coordination within the transportation system.
[68–69]	Wireless Sensor Networks (WSN)	WSN involves the use of sensors placed on or around vehicles to collect and transmit data. These sensors can monitor various parameters like road conditions, weather, and vehicle performance, enhancing the overall situational awareness.
[70–73]	Internet of Things (IoT)	CIoV leverages IoT technologies to connect vehicles and infrastructure to the internet. This connectivity enables the collection of data from various sources, fostering data-driven insights for transportation management.
[74–80]	5G Technology	The adoption of 5G networks enhances communication speed, reliability, and capacity. It supports the large-scale data transfer and low-latency communication required for real-time applications in CIoV.

(*Continued*)

TABLE 6.4 (*Continued*)
Summary of Communication Technologies Employed in CIoV

References	Technologies	Summary
[81–84]	Cloud computing	Cloud-based communication enables the storage and analysis of data gathered from vehicles, supporting real-time analytics, informed decision-making, and the delivery of services to both vehicles and users.
[85–89]	Edge computing	Edge computing reduces latency and accelerates response times by handling data in proximity to its source. In Cognitive Internet of Vehicles (CIoV) applications, it is employed to enable real-time decision-making.
[90–93]	M2M communication	M2M communication allows devices and vehicles to exchange information without human intervention. It is essential to CIoV's automation and intelligent decision-making processes.

The integration of these communication technologies forms the foundation of the CIoV, enabling a connected and intelligent transportation ecosystem.

In essence, the CIoV is a development of smart transportation systems in which automobiles have cognitive abilities to improve overall efficiency, safety, and user experience within the framework of modern transportation, in addition to communicating with the infrastructure and one another.

6.5 INTEGRATION OF 5G WITH FOG/EDGE COMPUTING IN CIoV

The future of intelligent transportation systems is greatly influenced by the convergence of edge computing, 5G, and the CIoV. Vehicle networks are made more intelligent, safe, and efficient by the distinct contributions made by each of these technologies.

The utilization of 5G technology on the IoV is discussed by the authors of Ref. [94] to facilitate the efficient transmission of real-time traffic information to drivers. Traditional methods, like employing macro base stations for processing offloaded vehicle applications, have become less effective due to the growing complexity of computing applications within vehicles. To address this challenge, the author introduces the concept of edge computing within 5G networks, pushing computing services to the network's edge. This approach is well-suited for executing IoV computing applications as it enables the offloading of applications to edge nodes (ENs) situated near vehicles. However, to prevent individual ENs from becoming overloaded, the author emphasizes the need to offload applications from overloaded ENs to other idle ENs, ensuring global load balance. By using multi-objective optimization to choose suitable destination ENs and efficiently manage computing task distribution, the proposed computation offloading approach for IoV, known as COV, seeks to optimize offloading latency and cost across ENs while attaining load balance. In summary, the authors leverage 5G technology in the IoV context, optimizing the processing and offloading of vehicle applications through edge computing and efficient resource management across Edge Nodes.

In Ref. [95], the study discusses the attainment of 5G and edge computing by deploying both micro and macro base stations, employing spectrum multiplexing, and utilizing cloud and fog computing architectures. It emphasizes the difficulties associated with managing mobility support, geographic distribution, and location awareness. Furthermore, it delves into the creation of fog computing architectures for vehicles and the effective deployment of tasks for multiple users. These innovations are geared towards enhancing sensitivity, interactivity, and the overall quality of experience within the CIoV environment.

In Ref. [96], the study revolves around the impact of 5G communication technology and fog computing in the context of the CIoV. Also, it underscores the supportive role of 5G technology in the integration of IoV and V2X. It proposes the adoption of fog computing to address handover issues among vehicles within the CIoV. Fog computing facilitates localized processing and storage of data, minimizing the necessity for extensive data transmission over long distances. These technological interventions are geared towards enhancing communication in vehicular ad-hoc networks (VANETs) and fostering trust among the interconnected entities in the CIoV ecosystem.

In Ref. [97], edge computing is a decentralized computing paradigm that, instead of depending on a centralized cloud server, places computation and data storage strategically close to the point of need. Its main goal is to reduce latency and improve connected devices' quality of service. In 5G-enabled VANETs, edge computing plays a crucial role by dispersing computer power close to the vehicles. This facilitates expedited data processing and elevates overall network efficiency. Such an approach proves instrumental in the collaborative distribution of vehicular content in 5G-VANETs, effectively addressing challenges related to resource optimization, network utilization, and the efficiency of data sharing.

In Ref. [98], the performance of vehicular ad-hoc networks (VANETs) is significantly boosted by the 5G network, offering reliable and low-latency communication services. The anticipated surge in internet-connected vehicles, projected by 2020, underscores the pivotal role of VANETs as an integral element of the IoV. The synergy of edge computing, also referred to as fog computing, with 5G, contributes to the efficient oversight and management of the entire network. This collaborative approach allows for the deployment of services and applications at the top of the network, optimizing routing and switching equipment to enhance traffic efficiency, thereby ensuring both passenger and street safety. Positioned as a complement to 5G networks, edge computing operates by pushing computing services to the periphery of the radio network [99].

In conclusion, the integration of edge computing and 5G within the framework of the CIoV results in a transportation system that is not only more responsive and intelligent but also more efficient. Together, these technologies play a crucial role in improving safety, minimizing latency, and facilitating instantaneous decision-making, setting the groundwork for the evolution of a smart and interconnected transportation ecosystem.

6.6 FUTURE DIRECTIONS

In the future, efforts will be made to explore and suggest advanced security mechanisms and protocols to tackle the inherent security challenges in CIoV architectures. This involves examining encryption techniques, authentication methods, and anomaly detection systems tailored to the unique characteristics of vehicular networks. Additionally, there will be a focus on investigating techniques to preserve user privacy while facilitating effective data sharing among vehicles. This may include developing privacy-preserving algorithms or employing methods such as differential privacy to protect sensitive information. Furthermore, research and design efforts will be directed towards creating user-friendly interfaces for CIoV systems. This aims to ensure that both drivers and pedestrians can easily interact with and comprehend the information provided by intelligent vehicular networks. Considerations will include exploring augmented reality interfaces and voice-based interactions to enhance user experience. The energy efficiency challenges associated with CIoV, particularly in the context of resource-constrained devices within vehicles, will be addressed. This involves exploring energy-efficient communication protocols, implementing adaptive power management strategies, and considering sustainable approaches for powering CIoV devices. Ethical implications of CIoV will be scrutinized, encompassing issues related to data ownership, accountability, and decision-making in critical situations. Frameworks for responsible AI use in vehicular networks will be proposed, considering potential societal impacts. By addressing future research directions, this chapter aims to contribute to the continuous development and comprehension of CIoV, fostering advancements in safety, efficiency, and the overall intelligence of vehicular networks.

6.7 CONCLUSION

In conclusion, this chapter has provided a comprehensive overview of the evolution, architectures, challenges, and advancements in CIoV, elucidating its role in connecting CIoT with vehicular networks. Throughout this exploration, we have highlighted the critical aspects of complexity, security imperatives, and communication technologies, underscoring the transformative potential of integrating edge computing and 5G technologies. A multimodal approach presents a promising way to maximize the effectiveness of CIoV and realize its potential to transform transportation. In order to handle the inherent complexity of cognitive systems in dynamic vehicular environments, firstly, researchers and practitioners should concentrate on creating robust and adaptive algorithms. Real-time processing ought to be given priority in these algorithms so that drivers can make decisions quickly and adapt to changing road conditions. Secondly, CIoV security is a critical issue that must be addressed by putting cutting-edge intrusion detection systems, authentication methods, and encryption techniques into place. Establishing comprehensive standards and regulations that guarantee the security and privacy of CIoV systems and the data they generate will require cooperation between academia, industry, and policymakers. Moreover, a strategic approach should be taken to the integration of edge computing and 5G technologies. Reducing latency and guaranteeing dependable connectivity require careful consideration of communication protocols and network architecture. Additionally, fostering collaboration among stakeholders in the development and deployment of 5G-enabled edge computing solutions will be instrumental in achieving a seamless and efficient CIoV ecosystem. In essence, this chapter serves as a concise guide for both researchers and practitioners, offering insights and recommendations to navigate the complexities of CIoV. By implementing the proposed solutions and fostering ongoing innovation, we can collectively work towards realizing the full potential of CIoV in shaping the future of intelligent transportation.

REFERENCES

1. Hudson, F.D. and Nichols, E.W., *2016*. The internet of things and cognitive computing. In *Handbook of Statistics* edited by V. N. Gudivada et al. (Elsevier B.V., Amsterdam, 2016), pp. 341–373.
2. Juniperresearch.com, *2015. 'Internet of Things' Connected Devices to Almost Triple to over 38 billion Units by 2020.* Available at: https://www.juniperresearch.com/press/pressreleases/iot-connected-devices-to-triple-to-38-bn-by-2020.
3. Singh, S., *2015*. Critical reasons for crashes investigated in the national motor vehicle crash causation survey (No. DOT HS 812 115).
4. Fagnant, D.J. and Kockelman, K., *2015*. Preparing a nation for autonomous vehicles: opportunities, barriers, and policy recommendations. *Transportation Research Part A: Policy and Practice*, *77*, pp. 167–181.
5. Savaglio, C. and Fortino, G., *2015*. Autonomic and cognitive architectures for the Internet of Things. In *Internet and Distributed Computing Systems: 8th International Conference, IDCS 2015*, Windsor, UK, September 2–4, 2015. Proceedings 8 (pp. 39–47). Springer International Publishing.
6. Delhi, S.I.N., *2016*. Automotive revolution & perspective towards 2030. *Auto Tech Review*, *4*(5), pp.20–25.
7. Laghari, A.A., Wu, K., Laghari, R.A., Ali, M. and Khan, A.A., *2021*. A review and state of art of Internet of Things (IoT). *Archives of Computational Methods in Engineering*, *29*, pp. 1–19.
8. Lombardi, M., Pascale, F. and Santaniello, D., *2021*. Internet of things: a general overview between architectures, protocols and applications. *Information*, *12*(2), p. 87.
9. Meneghello, F., Calore, M., Zucchetto, D., Polese, M. and Zanella, A., *2019*. IoT: internet of threats? A survey of practical security vulnerabilities in real IoT devices. *IEEE Internet of Things Journal*, *6*(5), pp. 8182–8201.
10. F. Al-Turjman and A. Radwan, *October 2017*. Data delivery in wireless multimedia sensor networks: challenging and defying in the IoT era. *IEEE Wireless Communications*, *24*(5), pp. 126–131 doi: 10.1109/WCM.2017.1700054.
11. Umamaheswari, S., Arun Kumar, S. and Sasikala, S., *2023*. Expert systems for improving the effectiveness of remote health monitoring in COVID-19 pandemic: a critical review. In Karrupusamy, P., Kalaiselvi, M., and Kumar, R. (Eds.), *System Design for Epidemics Using Machine Learning and Deep Learning* (pp.99–121). Wiley-Scrivener Publishing, Hoboken, NJ.
12. Vermesan, O., Eisenhauer, M., Sundmaeker, H., Guillemin, P., Serrano, M., Tragos, E.Z., Valiño, J., Gluhak, A. and Bahr, R., *2022*. Internet of things cognitive transformation technology research trends and applications. In *Cognitive Hyperconnected Digital Transformation* (pp. 17–95). River Publishers, Denmark.
13. Zhang, Y., Ma, X., Zhang, J., Hossain, M.S., Muhammad, G. and Amin, S.U., *2019*. Edge intelligence in the cognitive internet of things: improving sensitivity and interactivity. *IEEE Network*, *33*(3), pp. 58–64.

14. Ploennigs, J., Ba, A. and Barry, M., *2017.* Materializing the promises of cognitive IoT: how cognitive buildings are shaping the way. *IEEE Internet of Things Journal, 5*(4), pp.2367–2374.
15. Vlacheas, P., Giaffreda, R., Stavroulaki, V., Kelaidonis, D., Foteinos, V., Poulios, G., Demestichas, P., Somov, A., Biswas, A.R. and Moessner, K., *2013.* Enabling smart cities through a cognitive management framework for the internet of things. *IEEE Communications Magazine*, 51(6), pp. 102–111.
16. Li, F., Lam, K.Y., Li, X., Sheng, Z., Hua, J. and Wang, L., *2019.* Advances and emerging challenges in cognitive internet-of-things. *IEEE Transactions on Industrial Informatics, 16*(8), pp. 5489–5496.
17. Guerrero-Ibanez, J.A., Zeadally, S. and Contreras-Castillo, J., *2015.* Integration challenges of intelligent transportation systems with connected vehicle, cloud computing, and internet of things technologies. *IEEE Wireless Communications, 22*(6), pp. 122–128.
18. Guerrero-Ibáñez, J., Zeadally, S. and Contreras-Castillo, J., *2018.* Sensor technologies for intelligent transportation systems. *Sensors, 18*(4), p. 1212.
19. Mishra, R., Singh, A. and Kumar, R., *2016, March.* VANET security: issues, challenges and solutions. In *2016 International Conference on Electrical, Electronics, and Optimization Techniques (ICEEOT)* (pp. 1050–1055). IEEE, Chennai, India.
20. Contreras-Castillo, J., Zeadally, S. and Guerrero-Ibañez, J.A., *2017.* Internet of vehicles: architecture, protocols, and security. *IEEE internet of things Journal, 5*(5), pp. 3701–3709.
21. Cooperation, A.P.E., *2014.* White paper of internet of vehicles. In *Proceedings of the 50th Telecommunication and Information Working Group Meeting*, Brisbane, Australia (Vol. 29).
22. Lengton, M., Verzijl, D. and Dervojeda, K., *2015.* Internet of things: connected cars. In *Business Innovation Observatory. Business Innovation Observatory*, Brussels, Belgium.
23. Yang, F., Wang, S., Li, J., Liu, Z. and Sun, Q., *2014.* An overview of internet of vehicles. *China Communications, 11*(10), pp. 1–15.
24. Shuriya, B., Umamaheswari, S., Rajendran, A. and Sivaprakash, P., *2023, June.* One-dimensional dilated hypothesized learning method for intrusion detection system under constraint resource environment. In *2023 2nd International Conference on Advancements in Electrical, Electronics, Communication, Computing and Automation (ICAECA)* (pp. 1–6). IEEE, Coimbatore, India.
25. Qureshi, M.A., Noor, R.M., Shamshirband, S., Parveen, S., Shiraz, M. and Gani, A., *2015.* A survey on obstacle modeling patterns in radio propagation models for vehicular ad hoc networks. *Arabian Journal for Science and Engineering, 40*, pp. 1385–1407.
26. Dubey, B.B., Chauhan, N., Chand, N. and Awasthi, L.K., *2015.* Analyzing and reducing impact of dynamic obstacles in vehicular ad-hoc networks. *Wireless Networks, 21*, pp. 1631–1645.
27. Sommer, C., Joerer, S., Segata, M., Tonguz, O.K., Cigno, R.L. and Dressler, F., *2014.* How shadowing hurts vehicular communications and how dynamic beaconing can help. *IEEE Transactions on Mobile Computing, 14*(7), pp. 1411–1421.
28. Yan, T., Zhang, W. and Wang, G., *2014.* A grid-based on-road localization system in VANET with linear error propagation. *IEEE Transactions on Wireless Communications, 13*(2), pp. 861–875.
29. Alam, N., Balaei, A.T. and Dempster, A.G., *2012.* Relative positioning enhancement in VANETs: a tight integration approach. *IEEE Transactions on Intelligent Transportation Systems, 14*(1), pp. 47–55.
30. Yao, J., Balaei, A.T., Hassan, M., Alam, N. and Dempster, A.G., *2011.* Improving cooperative positioning for vehicular networks. *IEEE Transactions on Vehicular Technology, 60*(6), pp. 2810–2823.

31. Yang, F., Wang, S., Li, J., Liu, Z. and Sun, Q., *2014*. An overview of internet of vehicles. *China Communications*, *11*(10), pp. 1–15.
32. Corser, G.P., Fu, H. and Banihani, A., *2016*. Evaluating location privacy in vehicular communications and applications. *IEEE Transactions on Intelligent Transportation Systems*, *17*(9), pp. 2658–2667.
33. Huang, X., Yu, R., Kang, J., Wang, N., Maharjan, S. and Zhang, Y., *2016*. Software defined networking with pseudonym systems for secure vehicular clouds. *IEEE Access*, *4*, pp. 3522–3534.
34. Ying, B., Makrakis, D. and Mouftah, H.T., *2013*. Dynamic mix-zone for location privacy in vehicular networks. *IEEE Communications Letters*, *17*(8), pp. 1524–1527.
35. Monteiro, M.E.P., Rebelatto, J.L. and Souza, R.D., *2015*. Information-theoretic location verification system with directional antennas for vehicular networks. *IEEE Transactions on Intelligent Transportation Systems*, *17*(1), pp. 93–103.
36. Malandrino, F., Borgiattino, C., Casetti, C., Chiasserini, C.F., Fiore, M. and Sadao, R., *2014*. Verification and inference of positions in vehicular networks through anonymous beaconing. *IEEE Transactions on Mobile Computing*, *13*(10), pp. 2415–2428.
37. Fogue, M., Martinez, F.J., Garrido, P., Fiore, M., Chiasserini, C.F., Casetti, C., Cano, J.C., Calafate, C.T. and Manzoni, P., *2014*. Securing warning message dissemination in VANETs using cooperative neighbor position verification. *IEEE Transactions on Vehicular Technology*, *64*(6), pp. 2538–2550.
38. Karthikeyan, G., Yugesh, M., Umamaheswari, S., Sivashanmathi, A. and Kanna, K.N., *2023, June*. Integration of a state-of-the-art waste management tracking system. In *2023 2nd International Conference on Advancements in Electrical, Electronics, Communication, Computing and Automation (ICAECA)* (pp. 1–6). IEEE, Coimbatore, India.
39. Sharma, S. and Kaushik, B., *2019*. A survey on internet of vehicles: applications, security issues & solutions. *Vehicular Communications*, *20*, p. 100182.
40. Contreras, J., Zeadally, S. and Guerrero-Ibanez, J.A., *2017*. Internet of vehicles: architecture, protocols, and security. *IEEE Internet Things of Journal*, 99, pp. 1–9.
41. Fagnant, D.J. and Kockelman, K., *2015*. Preparing a nation for autonomous vehicles: opportunities, barriers and policy recommendations. *Transportation Research Part A: Policy and Practice*, *77*, pp. 167–181.
42. Umamaheswari, S., Arun Kumar, S. and Sasikala, S., *2023*. Expert systems for improving the effectiveness of remote health monitoring in COVID-19 pandemic: a critical review. In Karrupusamy, P., Kalaiselvi, M., and Kumar, R. (Eds.), *System Design for Epidemics Using Machine Learning and Deep Learning* (pp. 99–121). Wiley-Scrivener Publishing, Hoboken, NJ.
43. Kumari, S.M. and Geethanjali, N., *2010*. A survey on shortest path routing algorithms for public transport travel. *Global Journal of Computer Science and Technology*, *9*(5), pp. 73–76.
44. Pinggera, P., Ramos, S., Gehrig, S., Franke, U., Rother, C. and Mester, R., *2016, October*. Lost and found: detecting small road hazards for self-driving vehicles. In *2016 IEEE/RSJ International Conference on Intelligent Robots and Systems (IROS)* (pp. 1099–1106). IEEE, Daejeon, South Korea.
45. Lu, H., Liu, Q., Tian, D., Li, Y., Kim, H. and Serikawa, S., *2019*. The cognitive internet of vehicles for autonomous driving. *IEEE Network*, *33*(3), pp. 65–73.
46. Paul, A., Daniel, A., Ahmad, A. and Rho, S., *2015*. Cooperative cognitive intelligence for internet of vehicles. *IEEE Systems Journal*, *11*(3), pp. 1249–1258.
47. Rawat, D.B., Alsabet, R., Bajracharya, C. and Song, M., *2018*. On the performance of cognitive internet-of-vehicles with unlicensed user-mobility and licensed user-activity. *Computer Networks*, *137*, pp. 98–106.
48. Jadaan, K., Zeater, S. and Abukhalil, Y., *2017*. Connected vehicles: an innovative transport technology. *Procedia Engineering*, *187*, pp. 641–648.

49. Qian, Y., Jiang, Y., Hu, L., Hossain, M.S., Alrashoud, M. and Al-Hammadi, M., *2020.* Blockchain-based privacy-aware content caching in cognitive internet of vehicles. *IEEE network*, *34*(2), pp. 46–51.
50. Ding, N., Meng, X., Xia, W., Wu, D., Xu, L. and Chen, B., *2019.* Multivehicle coordinated lane change strategy in the roundabout under internet of vehicles based on game theory and cognitive computing. *IEEE Transactions on Industrial Informatics*, *16*(8), pp. 5435–5443.
51. Ding, N., Meng, X., Xia, W., Wu, D., Xu, L. and Chen, B., *2019.* Multivehicle coordinated lane change strategy in the roundabout under internet of vehicles based on game theory and cognitive computing. *IEEE Transactions on Industrial Informatics*, *16*(8), pp. 5435–5443.
52. Lu, N., Cheng, N., Zhang, N., Shen, X. and Mark, J.W., *2014.* Connected vehicles: solutions and challenges. *IEEE Internet of Things Journal*, *1*(4), pp. 289–299.
53. Sun, Y., Wu, L., Wu, S., Li, S., Zhang, T., Zhang, L., Xu, J. and Xiong, Y., *2015, October.* Security and privacy in the internet of vehicles. In *2015 International Conference on Identification, Information, and Knowledge in the Internet of Things (IIKI)* (pp. 116–121). IEEE, Beijing, China..
54. Hou, X., Li, Y., Chen, M., Wu, D., Jin, D. and Chen, S., *2016.* Vehicular fog computing: a viewpoint of vehicles as the infrastructures. *IEEE Transactions on Vehicular Technology*, 65(6), pp. 3860–3873.
55. Yan, G., Rawat, D.B. and Bista, B.B., *2012, July.* Towards secure vehicular clouds. In *2012 Sixth International Conference on Complex, Intelligent, and Software Intensive Systems* (pp. 370–375). IEEE, Palermo, Italy.
56. Kaiwartya, O., Abdullah, A.H., Cao, Y., Altameem, A., Prasad, M., Lin, C.T. and Liu, X., *2016.* Internet of vehicles: motivation, layered architecture, network model, challenges, and future aspects. *IEEE Access*, *4*, pp. 5356–5373.
57. Chen, M., Tian, Y., Fortino, G., Zhang, J. and Humar, I., *2018.* Cognitive internet of vehicles. *Computer Communications*, *120*, pp. 58–70.
58. Bhatia, J., Modi, Y., Tanwar, S. and Bhavsar, M., *2019.* Software defined vehicular networks: a comprehensive review. *International Journal of Communication Systems*, *32*(12), p. e4005.
59. Joy, J. and Gerla, M., *2017, July.* Internet of vehicles and autonomous connected car-privacy and security issues. In *2017 26th International Conference on Computer Communication and Networks (ICCCN)* (pp. 1–9). IEEE, Vancouver, Canada.
60. Othmane, L.B., Weffers, H., Mohamad, M.M. and Wolf, M., *2015.* A survey of security and privacy in connected vehicles. In *Wireless Sensor and Mobile Ad-Hoc Networks: Vehicular and Space Applications* (pp.217–247). Springer, New York.
61. Feng, S. and Haykin, S., *2019.* Cognitive risk control for anti-jamming V2V communications in autonomous vehicle networks. *IEEE Transactions on Vehicular Technology*, *68*(10), pp. 9920–9934
62. Chang, H. and Ning, N., *2021.* An intelligent multimode clustering mechanism using driving pattern recognition in cognitive internet of vehicles. *Sensors*, *21*(22), p. 7588.
63. Ouamna, H., Madini, Z. and Zouine, Y., *2021, May.* Optimization of a V2V communication in cognitive radio context. In *2021 7th International Conference on Optimization and Applications (ICOA)* (pp. 1–6). IEEE, Wolfenbüttel, Germany.
64. Kamal, M., Srivastava, G. and Tariq, M., *2020.* Blockchain-based lightweight and secured v2v communication in the internet of vehicles. *IEEE Transactions on Intelligent Transportation Systems*, *22*(7), pp. 3997–4004.
65. Umamaheswari, S. and Rohini, R., *2023, June.* A novel dual band slotted PIFA antenna for vehicle to vehicle and vehicle to infrastructure communication. In *2023 2nd International Conference on Advancements in Electrical, Electronics, Communication, Computing and Automation (ICAECA)* (pp. 1–5). IEEE, Coimbatore, India.

66. He, Y., Wang, D., Huang, F., Zhang, R., Gu, X. and Pan, J., *2023*. A V2I and V2V collaboration framework to support emergency communications in ABS-aided Internet of Vehicles. *IEEE Transactions on Green Communications and Networking, 7*(4), pp. 2038–2051.
67. Mekala, M.S., Dhiman, G., Patan, R., Kallam, S., Ramana, K., Yadav, K. and Alharbi, A.O., *2022*. Deep learning-influenced joint vehicle-to-infrastructure and vehicle-to-vehicle communication approach for internet of vehicles. *Expert Systems, 39*(5), p. e12815.
68. Wang, D., Zhang, Q., Liu, J. and Yao, D., *2019*. A novel QoS-awared grid routing protocol in the sensing layer of Internet of vehicles based on reinforcement learning. *IEEE Access, 7*, pp. 185730–185739.
69. Kumar, P.M., Manogaran, G., Sundarasekar, R., Chilamkurti, N. and Varatharajan, R., *2018*. Ant colony optimization algorithm with internet of vehicles for intelligent traffic control system. *Computer Networks, 144*, pp. 154–162.
70. Siegel, J.E., *2016*. *Data proxies, the cognitive layer, and application locality: enablers of cloud-connected vehicles and next-generation internet of things* (Doctoral dissertation, Massachusetts Institute of Technology).
71. Foteinos, V., Kelaidonis, D., Poulios, G., Vlacheas, P., Stavroulaki, V. and Demestichas, P., *2013*. Cognitive management for the internet of things: a framework for enabling autonomous applications. *IEEE Vehicular Technology Magazine, 8*(4), pp. 90–99.
72. Yao, Z., Wu, S. and Wen, Y., *2019*. Formation generation for multiple unmanned vehicles using multi-agent hybrid social cognitive optimization based on the internet of things. *Sensors, 19*(7), p. 1600.
73. Mahmood, Z., *2020*. *Connected Vehicles in the Internet of Things*. Springer Nature SwitzerlandAG, Cham, Switzerland.
74. Li, F., Lam, K.Y., Ni, Z., Niyato, D., Liu, X. and Wang, L., *2021*. Cognitive carrier resource optimization for internet-of-vehicles in 5G-enhanced smart cities. *IEEE Network, 36*(1), pp. 174–180.
75. Kombate, D., *2016, December*. The Internet of vehicles based on 5G communications. In *2016 IEEE International Conference on Internet of Things (iThings) and IEEE Green Computing and Communications (GreenCom) and IEEE Cyber, Physical and Social Computing (CPSCom) and IEEE Smart Data (SmartData)* (pp. 445–448). IEEE, Chengdu, China.
76. Mumtaz, S., Huq, K.M.S., Ashraf, M.I., Rodriguez, J., Monteiro, V. and Politis, C., *2015*. Cognitive vehicular communication for 5G. *IEEE Communications Magazine, 53*(7), pp. 109–117.
77. Almutairi, M.S., *2022*. Deep learning-based solutions for 5G network and 5G-enabled Internet of vehicles: advances, meta-data analysis, and future direction. *Mathematical Problems in Engineering, 2022*, pp. 1–27.
78. Wan, S., Gu, R., Umer, T., Salah, K. and Xu, X., *2020*. Toward offloading internet of vehicles applications in 5G networks. *IEEE Transactions on Intelligent Transportation Systems, 22*(7), pp. 4151–4159.
79. Ning, Z., Zhang, K., Wang, X., Obaidat, M.S., Guo, L., Hu, X., Hu, B., Guo, Y., Sadoun, B. and Kwok, R.Y., *2020*. Joint computing and caching in 5G-envisioned Internet of vehicles: a deep reinforcement learning-based traffic control system. *IEEE Transactions on Intelligent Transportation Systems, 22*(8), pp. 5201–5212.
80. Priya, K.H. and Umamaheswari, S., *2023, June*. Next generation optimized patch antenna for 5G applications. In *2023 2nd International Conference on Advancements in Electrical, Electronics, Communication, Computing and Automation (ICAECA)* (pp. 1–5). IEEE, Coimbatore, India..
81. Peter, D., Alavi, A., Javadi, B. and Fernandes, S.L. eds., *2020*. *The Cognitive Approach in Cloud Computing and Internet of Things Technologies for Surveillance Tracking Systems*. Academic Press, Cambridge, Massachusetts, USA.

82. Sharma, S., Awan, M.B. and Mohan, S., *2017, December.* Cloud enabled cognitive radio adhoc vehicular networking (CRAVENET) with security aware resource management and internet of vehicles (IoV) applications. In *2017 IEEE International Conference on Advanced Networks and Telecommunications Systems (ANTS)* (pp. 1–6). IEEE, Bhubaneswar, India.
83. Xu, X., Gu, R., Dai, F., Qi, L. and Wan, S., *2020.* Multi-objective computation offloading for internet of vehicles in cloud-edge computing. *Wireless Networks, 26*, pp. 1611–1629.
84. Chen, M. and Leung, V.C., *2018.* From cloud-based communications to cognition-based communications: a computing perspective. *Computer Communications, 128*, pp. 74–79.
85. Ning, Z., Zhang, K., Wang, X., Guo, L., Hu, X., Huang, J., Hu, B. and Kwok, R.Y., *2020.* Intelligent edge computing in internet of vehicles: a joint computation offloading and caching solution. *IEEE Transactions on Intelligent Transportation Systems, 22*(4), pp. 2212–2225.
86. Dai, Y., Xu, D., Maharjan, S., Qiao, G. and Zhang, Y., *2019.* Artificial intelligence empowered edge computing and caching for internet of vehicles. *IEEE Wireless Communications, 26*(3), pp. 12–18.
87. Wang, K., Wang, X., Liu, X. and Jolfaei, A., *2020.* Task offloading strategy based on reinforcement learning computing in edge computing architecture of internet of vehicles. *IEEE Access, 8*, pp. 173779–173789.
88. Zhao, J., Sun, X., Li, Q. and Ma, X., *2020.* Edge caching and computation management for real-time internet of vehicles: an online and distributed approach. *IEEE Transactions on Intelligent Transportation Systems, 22*(4), pp. 2183–2197.
89. Xu, X., Xue, Y., Qi, L., Yuan, Y., Zhang, X., Umer, T. and Wan, S., *2019.* An edge computing-enabled computation offloading method with privacy preservation for internet of connected vehicles. *Future Generation Computer Systems, 96*, pp. 89–100.
90. Tragos, E.Z. and Angelakis, V., *2013, June.* Cognitive radio inspired M2M communications. In *2013 16th International Symposium on Wireless Personal Multimedia Communications (WPMC)* (pp. 1–5). IEEE, Atlantic City, NJ, USA.
91. Ding, F., Su, R., Tong, E., Zhang, D., Zhu, H. and Ismail, M.W.M., *2018.* Toward a M2M-based Internet of vehicles framework for wireless monitoring applications. *IEEE Access, 6*, pp. 67699–67708.
92. Zhang, Y., Yu, R., Nekovee, M., Liu, Y., Xie, S. and Gjessing, S., *2012.* Cognitive machine-to-machine communications: visions and potentials for the smart grid. *IEEE Network, 26*(3), pp. 6–13.
93. Rawat, P., Singh, K.D. and Bonnin, J.M., *2016.* Cognitive radio for M2M and internet of things: a survey. *Computer Communications, 94*, pp. 1–29.
94. Wan, S., Li, X., Xue, Y., Lin, W. and Xu, X., *2020.* Efficient computation offloading for Internet of Vehicles in edge computing-assisted 5G networks. *The Journal of Supercomputing, 76*, pp. 2518–2547.
95. Zhou, P., Shen, K., Kumar, N., Zhang, Y., Hassan, M.M. and Hwang, K., *2020.* Communication-efficient offloading for mobile-edge computing in 5G heterogeneous networks. *IEEE Internet of Things Journal, 8*(13), pp. 10237–10247.
96. Gao, J., Agyekum, K.O.B.O., Sifah, E.B., Acheampong, K.N., Xia, Q., Du, X., Guizani, M. and Xia, H., *2019.* A blockchain-SDN-enabled Internet of vehicles environment for fog computing and 5G networks. *IEEE Internet of Things Journal, 7*(5), pp. 4278–4291.
97. Luo, G., Yuan, Q., Zhou, H., Cheng, N., Liu, Z., Yang, F. and Shen, X.S., *2018.* Cooperative vehicular content distribution in edge computing assisted 5G-VANET. *China communications, 15*(7), pp. 1–17.
98. Zhang, J., Zhong, H., Cui, J., Tian, M., Xu, Y. and Liu, L., *2020.* Edge computing-based privacy-preserving authentication framework and protocol for 5G-enabled vehicular networks. *IEEE Transactions on Vehicular Technology, 69*(7), pp. 7940–7954.
99. Feng, J., Liu, Z., Wu, C. and Ji, Y., *2017.* AVE: autonomous vehicular edge computing framework with ACO-based scheduling. *IEEE Transactions on Vehicular Technology, 66*(12), pp. 10660–10675.

Index

Note: **Bold** page numbers refer to tables and *italic* page numbers refer to figures.

FSC
www.fsc.org
MIX
Papier | Fördert gute Waldnutzung
FSC® C083411

Kolibri 360 GmbH
Ferdinand-Jühlke-Straße 7
99095 Erfurt, Deutschland
produktsicherheit@kolibri360.de